SpringerBriefs in Applied Sciences and Technology

SpringerBriefs present concise summaries of cutting-edge research and practical applications across a wide spectrum of fields. Featuring compact volumes of 50 to 125 pages, the series covers a range of content from professional to academic.

Typical publications can be:

- A timely report of state-of-the art methods
- An introduction to or a manual for the application of mathematical or computer techniques
- A bridge between new research results, as published in journal articles
- A snapshot of a hot or emerging topic
- An in-depth case study
- A presentation of core concepts that students must understand in order to make independent contributions

SpringerBriefs are characterized by fast, global electronic dissemination, standard publishing contracts, standardized manuscript preparation and formatting guidelines, and expedited production schedules.

On the one hand, **SpringerBriefs in Applied Sciences and Technology** are devoted to the publication of fundamentals and applications within the different classical engineering disciplines as well as in interdisciplinary fields that recently emerged between these areas. On the other hand, as the boundary separating fundamental research and applied technology is more and more dissolving, this series is particularly open to trans-disciplinary topics between fundamental science and engineering.

Indexed by EI-Compendex, SCOPUS and Springerlink.

Arnold A. Lubguban · Arnold C. Alguno ·
Roberto M. Malaluan · Gerard G. Dumancas

Computational Thermo-kinetics of Rigid Polyurethane Foams

Theory and Applications

 Springer

Arnold A. Lubguban
Department of Chemical Engineering
and Technology, Center for Sustainable
Polymers
Mindanao State University-Iligan Institute
of Technology
Iligan City, Lanao del Norte, Philippines

Roberto M. Malaluan
Department of Chemical Engineering
and Technology, Center for Sustainable
Polymers
Mindanao State University-Iligan Institute
of Technology
Iligan City, Lanao del Norte, Philippines

Arnold C. Alguno
Department of Materials Resources
Engineering and Technology, Center for
Sustainable Polymers
Mindanao State University-Iligan Institute
of Technology
Iligan City, Lanao del Norte, Philippines

Gerard G. Dumancas
Department of Chemistry
North Carolina Agricultural and Technical
State University
Greensboro, NC, USA

ISSN 2191-530X   ISSN 2191-5318 (electronic)
SpringerBriefs in Applied Sciences and Technology
ISBN 978-981-96-2076-0   ISBN 978-981-96-2077-7 (eBook)
https://doi.org/10.1007/978-981-96-2077-7

This book is dedicated to all the researchers, engineers, and scientists who relentlessly pursue innovation in materials science and sustainable engineering. May your dedication to knowledge and discovery continue to inspire generations to come.

To our mentors and educators, whose guidance shaped our paths and instilled in us a commitment to excellence in research and teaching. Your impact on our lives and careers is immeasurable.

And to our families, friends, and loved ones, for their unwavering support, patience, and encouragement throughout the many hours of study and writing. This achievement is as much yours as it is ours.

Foreword

The symbiotic relationship between empirical observation and theoretical insight is often central to the scientific discovery process. This relationship has changed dramatically in recent years, with computational modeling transcending its traditional role in physical sciences and becoming an indispensable tool across disciplines. This advancement is exemplified in Computational Thermo-kinetics for the Advancement of Rigid Polyurethane Foams: Theory and Applications by the investigation of Rigid Polyurethane Foams. This book demonstrates how computational thermo-kinetics can be used to design materials with unprecedented precision, performance, and sustainability, revealing a future for RPUFs that balance technical rigor with environmental responsibility.

Rigid polyurethane foams are a material of extraordinary utility and significance, with applications ranging from construction to automotive and aerospace. Their structure provides unique insulating, mechanical, and thermal properties, making them ideal for energy-efficient buildings, lightweight automotive parts, and a wide range of other applications. Traditionally, RPUF design has relied heavily on iterative experimentation, which is limited in time and resources, while exacerbating environmental impact. This book introduces readers to a novel approach: computational modeling and thermo-kinetics as tools for predicting and refining foam properties before physical trials, resulting in faster, more resource-efficient development cycles.

Beginning with foundational theories, it delves into the principles of predictive formulation and machine learning-driven material design, offering a framework for researchers to leverage advanced computational methods. In Computational Thermo-kinetics for the Advancement of Rigid Polyurethane Foams, each chapter builds upon the last, showing how computational models are reshaping the future of RPUFs. Chapters on material composition, mechanical and chemical properties, thermal stability, and environmental factors provide theoretical grounding and actionable insights. At the same time, dedicated sections on sustainability, recyclability, and regulatory compliance reflect a commitment to environmental stewardship.

As this book illustrates, the future of RPUF development lies in the intelligent integration of computational techniques with sustainable practices. Predictive models that minimize waste, formulations that replace petrochemical bases with renewable sources, and simulations that anticipate product lifecycles all point toward a more responsible approach to materials science.

With admiration for the dedication and expertise that brought this book to fruition, we are honored to introduce this work to its readers. I am confident that it will inspire future research, foster interdisciplinary collaboration, and contribute significantly to the advancement of rigid polyurethane foam technology and material science as a whole.

Scranton, USA   Riddhiman Medhi, Ph.D.

Preface

The ever-evolving field of material science has witnessed remarkable advancements over recent decades, yet few materials have shown as much promise and versatility as rigid polyurethane foams (RPUFs). Whether in construction, automotive, aerospace, or beyond, RPUFs have become integral to applications demanding lightweight structure, thermal insulation, and environmental resilience. Computational Thermo-kinetics for the Advancement of Rigid Polyurethane Foams: Theory and Applications emerges from the need for a deeper understanding of the complex behaviors and vast potential of these foams, particularly as they intersect with the transformative power of computational modeling and thermo-kinetics.

The inspiration for this book is rooted in the combined years of research and experience with computational thermo-kinetics, where we witnessed firsthand how computational tools could transform RPUF research and application. With advancements in machine learning, predictive formulation, and multiscale modeling, we can now reduce the experimental timeframes, resource consumption, and environmental impact traditionally associated with material development. This book seeks to explore these advancements while offering a comprehensive roadmap for researchers, engineers, and students engaged in the study and application of RPUFs.

Designed to serve as both a technical resource and a forward-looking guide, the book begins by establishing the theoretical understanding of RPUFs, providing readers with a foundational knowledge of their chemical, mechanical, and thermal properties. Building on this, we delve into the practical aspects of computational modeling, machine learning-driven predictions, and predictive formulation techniques, each explained in the context of their transformative impact on the field. To support thorough exploration, each chapter is accompanied by case studies, data analyses, and references to recent advancements, bridging theory with application.

This work also prioritizes sustainability—an imperative in today's material science landscape. Recognizing the environmental pressures facing industries worldwide, the book addresses the role of bio-based components, recyclability, and environmental performance standards within RPUF development. We aim to highlight how the combination of computational modeling with sustainable practices can lead to RPUFs that meet performance demands without compromising ecological responsibility.

The journey of writing this book has been both challenging and rewarding, requiring extensive research, collaboration, and reflection. We are grateful for the contributions of our colleagues, mentors, and students, whose insights and innovations have profoundly shaped this work. We especially appreciate the support from the professional community, who have continually inspired us to explore new frontiers in material science.

It is our hope that this book not only serves as a technical reference but also ignites curiosity and inspires further research into the limitless possibilities of RPUFs. The advancements shared in these pages are intended to benefit both academic research and practical industry application, promoting a future where material science thrives in synergy with computational technology and environmental stewardship.

Iligan City, Philippines   Arnold A. Lubguban
Iligan City, Philippines   Arnold C. Alguno
Iligan City, Philippines   Roberto M. Malaluan
Greensboro, USA   Gerard G. Dumancas

Acknowledgements

The successful completion of this multiauthored work, Computational Thermo-kinetics for the Advancement of Rigid Polyurethane Foams: Theory and Applications, would not have been possible without the support, guidance, and collaboration of numerous individuals and institutions. We deeply thank everyone who contributed their expertise, resources, and time to this project.

We extend our deepest gratitude to the *Department of Science and Technology—Philippine Council for Industry, Energy and Emerging Technology Research and Development (DOST-PCIEERD)* for their invaluable support in advancing research and innovation in computational thermo-kinetics. Their funding, technical assistance, and commitment to fostering scientific progress have been instrumental in realizing the objectives of this work. We are profoundly thankful to *Mindanao State University-Iligan Institute of Technology* for its unwavering institutional support and commitment to advancing research in sustainable and innovative materials. The resources provided by MSU-IIT were invaluable to the comprehensive research and analysis presented in this book. Special thanks to the faculty and administrative staff who facilitated our work and made their expertise available to us at every stage.

Our sincere appreciation also goes to the *Center for Sustainable Polymers*, whose dedication to the development of eco-friendly materials has inspired much of the work found within these pages. The Center's state-of-the-art facilities, generous research grants, and collaborative spirit allowed us to explore new horizons in sustainable and high-performance polyurethane foams. We are especially grateful to the research team at the Center, whose insights and technical guidance enriched the rigor and quality of our research.

Our deepest appreciation to our main editor and author, Arnold A. Lubguban Ph.D., for his invaluable time and effort in shaping this book into the best it could be. Additionally, Roberto M. Malaluan DoE., Arnold C. Alguno Ph.D., and Gerard G. Dumancas Ph.D., for their equally great contributions to the book as editors. This book would not have been possible without the dedication of our co-authors, research assistants, and data analysts, who worked tirelessly to contribute their specialized knowledge to this project. Special thanks to Roger Jr. G. Dingcong (Chap. 2), Shem Patrick Josh A Laguda (Chaps. 1–4), Fortia Louise Adeliene M. Alfeche (Chap. 3),

David Jeiel G. Dabodabo (Chap. 2), Carlo James H. Solidor (Chap. 1), Mae Francy T. Lucid (Chap. 4), Glady May A. Mejias (Chap. 1), Jady Lee E. Amarillas (Chaps. 2 and 3), and to Jhon Alter A. Ortiz (Multimedia). To our editors, proofreaders, illustrators, and technical reviewers, thank you for your meticulous work and commitment to excellence.

Last, we are grateful to our families, friends, and mentors for their encouragement and support throughout this journey. Their patience and understanding allowed us to dedicate ourselves to this endeavor, and for that, we are truly thankful.

Contents

Contributors

Chapters Co-authors and Contributors

Jady Lee E. Amarillas Center for Sustainable Polymers, Mindanao State University-Iligan Institute of Technology, Iligan City, Lanao del Norte, Philippines, e-mail: jadylee.amarillas@g.msuiit.edu.ph

Fortia Louise Adeliene M. Alfeche Sustainable Resource Engineering Research for Construction Materials and Technology, Mindanao State University-Iligan Institute of Technology, Iligan City, Lanao del Norte, Philippines, e-mail: fortia.alfeche@g.msuiit.edu.ph

David Jeiel G. Dabodabo Center for Sustainable Polymers, Mindanao State University-Iligan Institute of Technology, Iligan City, Lanao del Norte, Philippines, e-mail: davidjeiel.dabodabo@g.msuiit.edu.ph

Roger G. Dingcong Jr. Center for Sustainable Polymers, Mindanao State University-Iligan Institute of Technology, Iligan City, Lanao del Norte, Philippines, e-mail: rogerjr.dingcong@g.msuiit.edu.ph

Shem Patrick Josh A. Laguda Center for Sustainable Polymers, Mindanao State University-Iligan Institute of Technology, Iligan City, Lanao del Norte, Philippines, e-mail: shempatrickjosh.laguda@g.msuiit.edu.ph

Mae Francy T. Lucid Center for Sustainable Polymers, Mindanao State University-Iligan Institute of Technology, Iligan City, Lanao del Norte, Philippines, e-mail: maefrancy.lucid@g.msuiit.edu.ph

Glady May A. Mejias Center for Sustainable Polymers, Mindanao State University-Iligan Institute of Technology, Iligan City, Lanao del Norte, Philippines, e-mail: gladymay.mejias@g.msuiit.edu.ph

Carlo James H. Solidor Center for Sustainable Polymers, Mindanao State University-Iligan Institute of Technology, Iligan City, Lanao del Norte, Philippines, e-mail: carlojames.solidor@g.msuiit.edu.ph

Abbreviations

2D	Two-Dimensional
AFLOWLIB	Automatic Flow for Materials Discovery
AI	Artificial Intelligence
ANN	Artificial Neural Networks
API	Artificial Programming Interfaces
ASTM	American Standards for Testing and Materials
BAML	Bonds, Angles, Machine Learning
CAFE	Corporate Average Fuel Economy
CFC	Chlorofluorocarbons
CFD	Computational Fluid Dynamics
COD	Crystallography Open Database
CSD	Cambridge Structural Database
DBTDL	Dibutyltin Dilaurate
EO	Ethylene Oxide
EPD	Environmental Product Declarations
FAIR	Findable, Accessible, Interoperable, and Reusable
FEA	Finite Element Analysis
FEM	Finite Element Modeling
FVM	Finite Volume Method
GBR	Gradient Boosting Regression
GNN	Graph Neural Network
HCFC	Hydrochlorofluorocarbons
HFO	Hydrofluoroolefins
HOMO	Highest Occupied Molecular Orbital
ICDD	International Center for Diffraction Data
ICSD	Inorganic Crystal Structure Database
KRR	Kernel Ridge Regression
LEED	Leadership in Energy and Environmental Design
LR	Linear Regression
LUMO	Lowest Unoccupied Molecular Orbital
MAE	Mean Absolute Error

MD	Molecular Dynamics
MDI	Methylene Diphenyl Diisocyanate
ML	Machine Learning
NMR	Nuclear Magnetic Resonance
ODE	Ordinary Differential Equations
ODS	Ozone-Depleting Substances
OHV	Hydroxyl Value
OQMD	Open Quantum Materials Database
PBA	Physical Blowing Agents
PCM	Phase Change Materials
PDMS	Polydimethylsiloxane
PEO-PPO	Polyethylene oxide-co-propylene oxide
PO	Propylene Oxide
PU	Polyurethanes
QMSPR	Quantitative Materials Structure-Property Relationship
RH	Rice Husk
RMSE	Root Mean Square Error
RPUF	Rigid Polyurethane Foams
SEM	Scanning Electron Microscope
SMILES	Simplified Molecular Input Line Entry System
TDI	Toluene Diisocyanate
TPU	Thermoplastic Polyurethanes
UV	Ultraviolet

List of Figures

List of Tables

Chapter 1
Foundations of Computational Thermo-kinetics in RPUF

1.1 Overview of Rigid Polyurethane Foams

Polyurethanes. Polyurethanes (PU), in general, are a diverse class of materials with a wide variety of applications. They are used in coatings, adhesives, sealants, and elastomers (CASE applications) [1]. Still, these represent a small part of the usage of PUs. The most widely used variety of these materials is in foams, specifically rigid and flexible foams [2]. Rigid foams are closed-cell materials with excellent heat resistance when blown with the proper blowing agents (explained in Sect. 1.1.2). Rigid polyurethane foams represent a highly versatile class of materials with exceptional properties that make them indispensable in multiple industries. Unlike flexible foams, RPFs are designed to provide structural rigidity and superior insulation, achieving this through a cellular structure that encapsulates gas within closed cells, minimizing thermal conductivity [3]. This cellular architecture, with void volumes between 95 and 97% and cell sizes ranging from 100 to 500 μm, exhibits insulation properties by creating a low-density and low-thermal conductivity material. Figure 1.1 shows some applications of rigid PU foams. Comparatively, other insulation materials like expanded polystyrene, mineral wool, and even concrete require much greater thicknesses to achieve similar thermal efficiency [4, 5].

RPUFs are primarily created through a reaction between polyols and isocyanates. Polyols used in PU foam production often incorporate functional groups that influence the foam's mechanical and thermal properties [7]. For instance, aromatic polyester polyols are usually used for rigid polyurethane applications due to their economic viability and enhanced flame retardancy due to their aromatic backbones. These are widely used for applications that require thermal stability, such as in the construction industry where meeting stringent fire safety regulations is important for structural integrity and hazard mitigation on the buildings [5].

Polyols are unique because they are versatile materials that can be carefully selected and formulated based on desired attributes, such as rigidity, strength, and adhesive properties [1]. These properties are essential in applications like appliance

© The Author(s), under exclusive license to Springer Nature Singapore Pte Ltd. 2025
A. A. Lubguban et al., *Computational Thermo-kinetics of Rigid Polyurethane Foams*,
SpringerBriefs in Applied Sciences and Technology,
https://doi.org/10.1007/978-981-96-2077-7_1

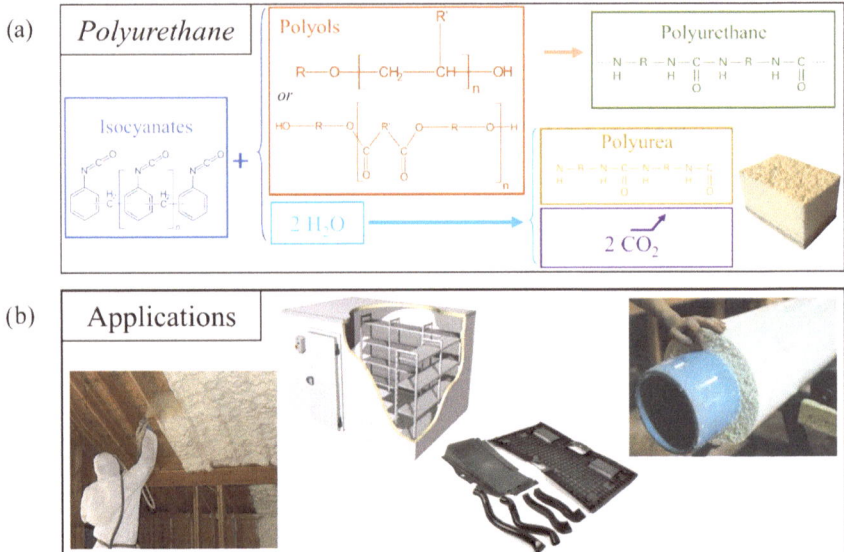

Fig. 1.1 Synthesis of rigid polyurethane foams and their applications. Reproduced from Borrero-López et al. [6]. Under the CC BY 4.0 license

insulation or sandwich panels, where RPFs act as insulators and structural adhesives [5].

Environmental concerns have spurred recent advancements toward more sustainable RPF formulations. Manufacturers are now exploring bio-based polyols derived from renewable resources like coconut, castor, and soybean. These bio-based polyols can reduce the carbon footprint of foam production while providing properties comparable to or exceeding those of petroleum-derived counterparts. For example, coconut oil-derived polyols align with sustainability goals and offer chemical stability that can be beneficial in foam applications. The shift to bio-based materials is promising, reflecting an industry-wide commitment to reducing environmental impact and enhancing product lifecycle sustainability.

As the industry advances, innovations in RPUF formulations and production methods continue to emerge, focusing on enhanced performance, sustainability, and adaptability to diverse applications. Subsequent sections will delve into these advancements, examining the chemical composition, mechanical properties, applications, and processing technologies that define the future of rigid polyurethane foams. In this chapter, we will discuss an overview of rigid PU foams, including their structure, synthesis, and history, how they came to be, and how they are currently meeting commercial and industrial demands.

Structure. Polyurethanes are versatile polymers defined by their carbamate or urethane linkages (Fig. 1.1) formed by the reaction between diisocyanates and polyols (polymers with multiple hydroxyl groups) [2, 7]. Figure 1.2 shows that the basic structure of polyurethane can be modified by selecting different types of isocyanates

Fig. 1.2 A urethane bond

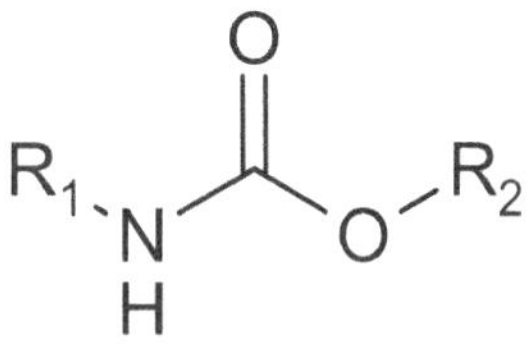

and polyols, resulting in a wide range of properties from highly flexible foams to tough, rigid, and lightweight material [2, 8] (Fig. 1.2).

The backbone of polyurethanes consists of alternating hard and soft segments, a key feature that governs their mechanical and physical properties. Diisocyanates, which contain two or more isocyanate (–NCO) groups, form the hard segments that impart rigidity and stability, while polyols provide the soft segments that contribute to elasticity. This segmentation allows polyurethanes to achieve unique phase separation within the polymer, giving rise to materials that can be both flexible and resilient. The combination of these hard and soft segments results in a microphase-separated morphology, where the hard segments act as reinforcing domains within the soft matrix [2, 7].

Molecular weight and the degree of crosslinking play an important role in determining polyurethanes' rigidity) and thermal stability. Higher molecular weight polyol typically has longer chains, resulting in the increase of its soft segments, which increases tensile strength, elongation, and glass transition temperatures (Tg), resulting in more flexible foams while crosslinking agents such as triols introduce rigidity by creating a three-dimensional network that limits polymer chain mobility [9, 10], resulting in rigid foams. This crosslinked structure is especially desirable for applications that require high load-bearing capacity and durability such as those in the construction industry, as it enhances the polymer's resistance to deformation and thermal breakdown [2].

Hydrogen bonding within the hard segments further enhances the structural integrity of polyurethanes, impacting their mechanical properties and thermal behavior. The strong hydrogen bonds between hard segments improve cohesion and contribute to the polymer's tensile strength [11] and thermal stability [12]. These hydrogen bonds can also influence the processing characteristics, making polyurethanes suitable for high-performance applications in environments with extreme temperature variations [2, 10].

PUFs can also be classified into open-cell and closed-cell. Open-cell polyurethane foams are characterized by a structure with interconnected cells, allowing air to move freely throughout the material. This open structure contributes to their softer and more flexible nature, making them suitable for applications that require cushioning, such as furniture and bedding. Because of the air-filled spaces, open-cell foams have a lower density than closed-cell foams, making them relatively lightweight. However, they tend to retain moisture more readily, limiting their suitability in areas exposed to high humidity or where moisture resistance is necessary [13, 14]. Figure 1.3 shows

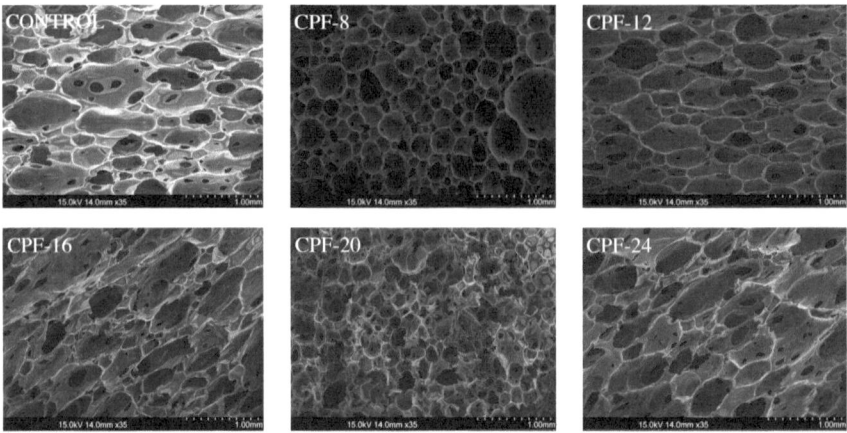

Fig. 1.3 SEM images of open-celled polyurethane foams derived from coconut monoglycerides and reproduced from Omisol et al. [15]. Under the CC BY 4.0 license

an example of the SEM images of open-celled polyurethane foams derived from coconut monoglycerides.

In contrast, closed-cell polyurethane foams consist of completely enclosed cells, trapping gas inside and making them more rigid and dense. This closed-cell structure enhances its thermal insulation properties, as the trapped gas provides effective resistance to heat flow [16]. Closed-cell foams are highly durable and resistant to water absorption, making them suitable for outdoor and high-moisture environments, such as in roofing insulation or flotation devices. Figure 1.4 shows scanning electron microscope (SEM) images of closed-cell polyurethane foams, highlighting their cellular morphology. Images (a) through (d) illustrate variations in cell size, wall thickness, and uniformity, which influence mechanical strength, thermal insulation, and water resistance. The enclosed cell structures prevent gas escape, contributing to superior insulation properties. These microstructural differences are key in optimizing foam performance for specific applications, such as load-bearing structures or moisture-resistant barriers. These foams are often more robust under compression and load-bearing applications due to their rigidity and higher density. As a result, closed-cell foams provide better structural support and stability, although they tend to be heavier and less flexible compared to open-cell foams [17] (Fig. 1.4).

Polyurethane structures can also vary between thermoplastic and thermosetting forms. Thermoplastic polyurethanes (TPUs) exhibit a linear, non-crosslinked molecular structure that allows them to be reshaped upon heating. [19, 20]. On the other hand, thermoset polyurethanes, formed with higher degrees of crosslinking, resist thermal degradation and solvent exposure, making them well-suited for demanding applications where stability under stress and temperature extremes is essential [2].

Ultimately, the properties of polyurethanes are closely tied to the ratio and type of hard and soft segments within their structure. The balance between these segments allows polyurethane materials to be tailored for specific uses, requiring

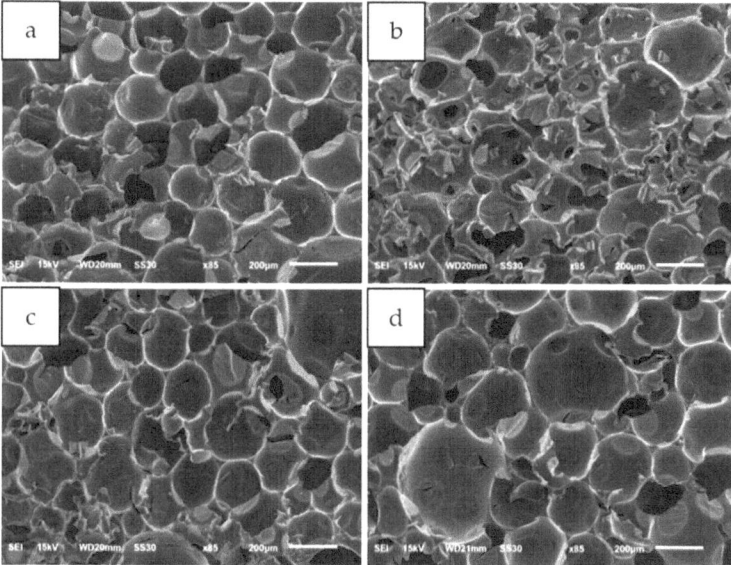

Fig. 1.4 SEM images of closed-cell foams derived from coconut oil. Reproduced from Alfeche et al. [18] under the CC BY 4.0 license

flexibility, rigidity, chemical resistance, or thermal stability. This adaptability makes polyurethanes a popular choice for various industries, from coatings and adhesives to medical devices and automotive components, as they can be engineered to meet a vast array of functional requirements [2, 9].

1.1.1 Synthesis

Polyurethanes (PUs) are highly adaptable polymers formed primarily through the polyaddition reaction of diisocyanates and polyols. First synthesized in the 1930s by Otto Bayer (more in the next section), PUs have since become central to various industries due to their remarkable versatility and capacity for customization. They are found in products ranging from flexible foams used in furniture and mattresses to rigid coatings and structural components in construction and automotive manufacturing [1]. The ability to tailor PU properties by selecting specific monomers, catalysts, and processing conditions allows manufacturers to create materials that meet diverse mechanical, thermal, and chemical requirements. With increasing demand for sustainable materials, research into PU synthesis has expanded beyond conventional methods, embracing innovations aimed at improving environmental safety and reducing reliance on petrochemical resources [21].

In conventional PU synthesis, a step-growth polymerization reaction takes place between a diisocyanate monomer and a polyol, usually a polyether or polyester. This

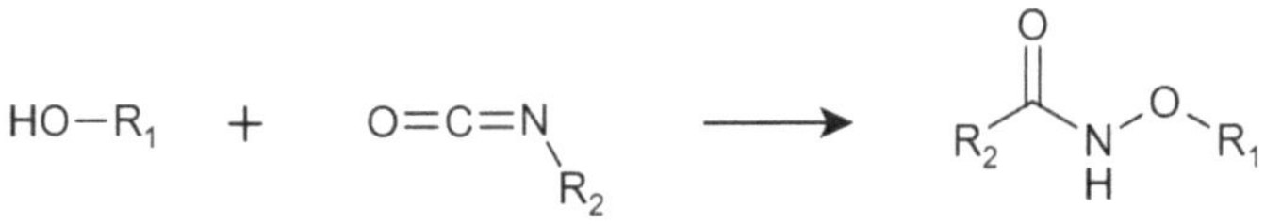

Fig. 1.5 Reaction between a polyol and an isocyanate to create a urethane bond

mechanism is called the gelling reaction. Common diisocyanates, such as toluene diisocyanate (TDI) and methylene diphenyl diisocyanate (MDI), react with polyols to form urethane linkages [22]. The choice of polyol significantly influences the properties of the resulting PU. Polyether polyols, for example, yield flexible and resilient polyurethanes, making them ideal for applications where elasticity is essential [6]. Although polyether polyols can still be used for RPUFs, they are usually tailored to be low in molecular weights with hydroxyl groups ranging from 3 to 8 per mol [22]. Figure 1.5 shows the reaction between a polyol and an isocyanate to form a urethane bond.

The traditional PU synthesis process can occur in either a single-stage reaction or a multistage process, depending on the desired properties and application of the final material. In a single-stage process, all monomers and catalysts are combined in one reaction vessel, resulting in a straightforward synthesis method. However, this approach provides less control over molecular weight and the polymer's mechanical properties. A multistage process is often preferred for applications requiring specific physical characteristics, such as high tensile strength or controlled elasticity. In the multistage method, a prepolymer is created by reacting an excess of diisocyanate with a polyol, yielding an isocyanate-terminated intermediate (Fig. 1.6). This prepolymer is then chain-extended with a diol, which increases the molecular weight and enhances the polymer's structural integrity. This two-step process allows for precise control over the final PU's mechanical and thermal properties, making it particularly suitable for high-performance applications in industries such as automotive and aerospace [21].

In the case of PU foams, two types of reaction simultaneously occur: the gelling action and blowing mechanism. The previous paragraph discussed the gelling process, which is the crosslinking and reaction of the polyol and diisocyanate. However, the urethane bonds could react further depending on certain circumstances, as found in Fig. 1.7. When there is an excess of isocyanates, reaction conditions, temperature, and catalyst presence, they could react to form allophanates but can

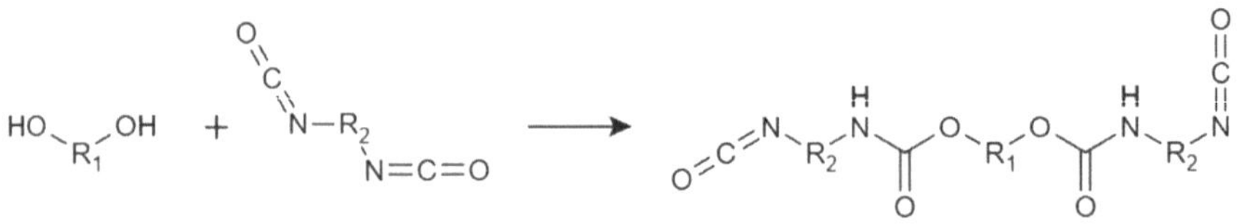

Fig. 1.6 Synthesis of the isocyanate-terminated prepolymer

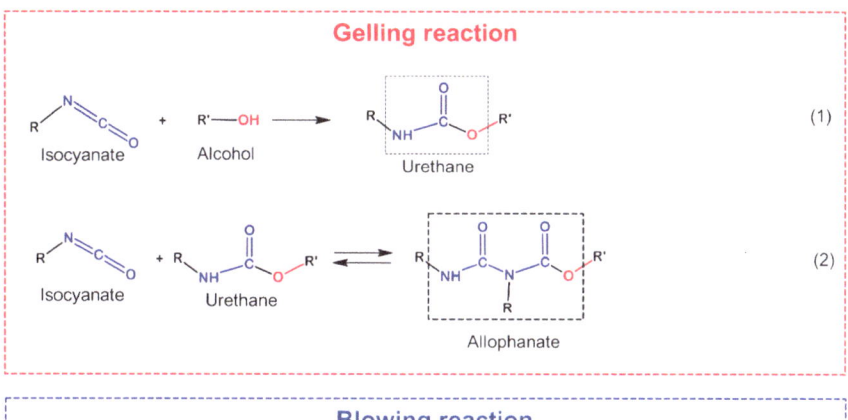

Fig. 1.7 Blowing and gelling reactions reproduced from Bondaug et al. [27]. Reproduced under CC BY 4.0 license

dissociate back into urethanes and isocyanates when it is subjected to high temperatures. In the case where water is used as a blowing agent, water reacts with isocyanates to create carbon dioxide and amine-terminated molecules. The amine reacts further with isocyanates to form urea bonds and reacts further to form biuret [23].

There are two types of blowing processes, either physical or chemical. In rigid polyurethane (PU) foam production, choosing the right blowing method is essential for creating a solid yet lightweight cellular structure. Physical blowing agents, like pentanes, vaporize under the heat of the polymeric reaction; this causes a blowing mechanism that creates the foam's cellular structure [22], which is tightly packed and provides excellent insulation properties and moisture resistance. These physical agents are particularly effective for rigid PU foams, as their rapid vaporization creates a small, uniform cell structure that enhances the foam's ability to minimize heat transfer and maintain structural stability under pressure [24, 25]. Originally, chlorofluorocarbons (CFCs) like CFC-11 and hydrofluorocarbons were used due to their influence on the low thermal conductivity of the RPUFs [3], but these are slowly phased out due to the implementation of the Montreal Protocol and other new policies which bans the use of ozone-depleting substances (ODSs) [26].

Chemical blowing agents, such as water, expand rigid PU foams through a different mechanism, producing gas via a chemical reaction with isocyanates that form carbon dioxide. This reaction is exothermic, adding heat to the system and gradually generating CO_2 gas, which helps to expand the foam structure. In rigid foams, this approach can result in larger cells compared to physical blowing agents alone, but it also creates a stable closed-cell structure. The carbon dioxide produced by water as a blowing agent leads to a high-density, durable foam suitable for applications like insulation panels, refrigeration units, and construction materials where both strength and low thermal conductivity are required [21].

For the gelling reaction, tin-based compounds, particularly dibutyltin dilaurate, are traditionally used to catalyze the PU-forming reaction. Just like other chemical reactions, the presence of a catalyst when polyurethanes are being synthesized is vital to obtain the desired properties. These highly efficient catalysts enable the rapid formation of urethane bonds by coordinating with the isocyanate group, enhancing its reactivity with the hydroxyl groups of polyols [3]. However, tin-based catalysts are associated with notable environmental and health concerns, especially in applications involving biomedical or food-contact materials where residual catalyst traces can pose toxicity risks. Removing these metal catalysts from the final polymer can be complex and costly, and any remaining traces may compromise the dielectric, mechanical, or biocompatibility properties of the PU product. In response to these challenges, the industry is actively pursuing alternative catalysis methods that retain high catalytic efficiency without adverse environmental or health impacts [28]. Some examples of catalysts can be found in Fig. 1.8.

In recent advancements within polyurethane (PU) foam synthesis, amine-based catalysts have gained significant attention due to their specialized ability to influence the precise kinetics of both gelling and blowing reactions (Handbook of Polymeric foams). Two key properties—basicity and steric accessibility—are important to a catalyst's effectiveness. Tertiary amines with higher basicity and minimal steric hindrance are more catalytically active, as they readily interact with isocyanate groups, forming complexes that promote the urethane-forming reaction with polyols or water. This interaction leads to efficient urethane linkages and the production of CO_2, which causes the blowing reaction in the foam synthesis [29].

Recently, studies on isocyanate-free PU synthesis have been extensively studied. As mentioned previously, the materials used in the synthesis of isocyanate-based polyurethanes can be highly toxic when inhaled or ingested by humans. Some innovative research has explored the potential of using carbon dioxide as a feedstock in isocyanate-free synthesis, converting this greenhouse gas into valuable polymer components. By incorporating renewable or recycled materials, isocyanate-free PUs support the development of carbon–neutral or carbon-negative materials, further improving the sustainability profile of polyurethane products [21, 28].

The industrial implications of these advancements in PU synthesis are significant, as they respond to the growing demand for high-performance and environmentally

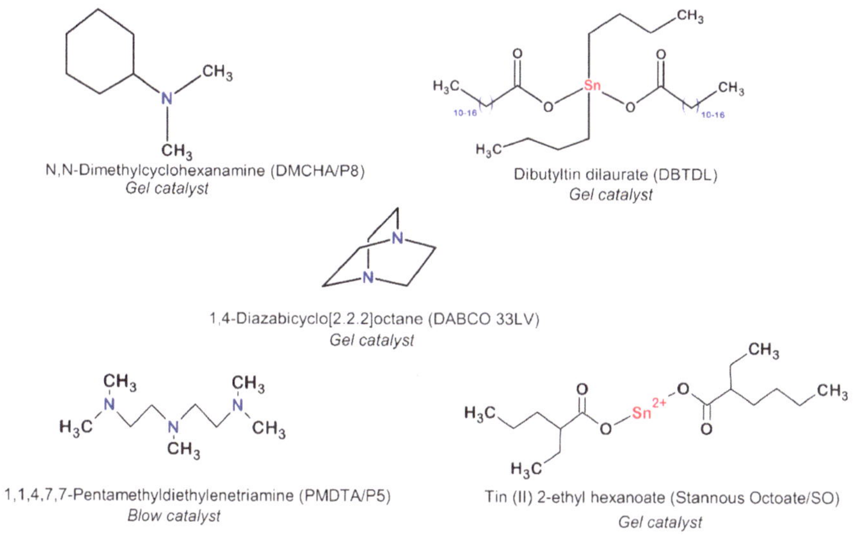

Fig. 1.8 Examples of gelling and blowing catalysts for polyurethane production reproduced from Bondaug et al. [27]. Reproduced under CC BY 4.0 01

responsible materials. Industries such as automotive, construction, and consumer goods are increasingly constrained by regulations and standards requiring safer chemicals and lower emissions. [21, 28].

Over the years, the synthesis of polyurethanes has evolved significantly, with a marked shift from traditional tin-catalyzed methods to greener alternatives that prioritize environmental safety and product performance. As research into these green synthesis techniques progresses, polyurethanes are poised to play a key role in the shift toward a sustainable future, providing versatile, high-performance materials for a wide range of applications.

1.1.2 Brief History and Timeline

The history of urethanes dates back to the 1800s when chemists Charles-Adolphe Wurtz and August Wilhelm von Hofmann first discovered the reaction between an isocyanate and a hydroxyl compound. While this reaction was groundbreaking, it would take several more decades before it found widespread application. It wasn't until the 1930s and 1940s, with advancements by Otto Bayer and his colleagues, that polyurethanes began to be synthesized on an industrial scale. Among these, the rigid polyurethane foams were one of the biggest interests at this time [3]. The onset of World War II amplified its use, and these foams became critical in military applications. Their unique properties made them ideal for insulating tanks, submarines, and aircraft. Their lightweight structure not only provided excellent thermal insulation but also offered durability and mechanical strength. Its applications were not really among the general public until later. In Germany, they gained moderate commercial acceptance, but only after the war did rigid polyurethane foams begin attracting attention in the United States. During 1945–1947, scientific investigations into German advancements led to detailed PB Reports on polyurethane systems and their applications, encouraging the U.S. Air Force to contract companies to research these materials [30].

After seeing the potential of these foams on the commercial scale, improvements were constantly made to create even more efficient and versatile foam formulations. The work of Bayer involved only the use of polyester polyols, and in the late 1950s, polyether polyols were introduced. Polyether polyols became the widely used material for RPUFs. Because of its economic viability, polyether polyols derived from polypropylene oxide shifted the industry into a new light. From there, the polyurethane foam industry experienced rapid growth and transformation. The introduction of polyether polyols in the late 1950s marked a turning point, as these materials offered better hydrolytic stability, improved flexibility, and lower cost compared to polyester polyols. With polyether polyols derived from polypropylene oxide, RPUFs became more affordable and accessible, opening up new applications in building insulation, appliance manufacturing, and automotive design [3].

The applications of RPUFs have expanded into a wide range of industries due to their exceptional insulation properties, structural integrity, and versatility. Today, RPUFs are extensively used in construction for thermal insulation in walls, roofs, and floors, enhancing energy efficiency in buildings. They are also essential in refrigeration for insulating appliances such as refrigerators and freezers, helping to maintain low temperatures while conserving energy [22]. Beyond these, RPUFs are applied in pipeline insulation, packaging, and even protective gear, highlighting their importance in the modern world.

Recently, with the depletion of its raw material from petroleum, there has been a need to shift to bio-based materials. This shift has driven significant research and innovation toward developing bio-based polyols derived from renewable resources, such as vegetable oils and agricultural by-products. These bio-based materials not only help reduce dependency on fossil fuels but also offer the potential for lower environmental impact throughout their lifecycle. The incorporation of bio-based components into RPUFs has led to the creation of sustainable alternatives that maintain the desirable properties of traditional polyurethane foams while contributing to a circular economy.

As consumer demand for environmentally friendly products grows, the industry is increasingly focused on optimizing formulations and production processes to enhance the performance and commercial viability of bio-based RPUFs, and now, with the emerging technology of computer simulation and AI, it is possible to predict the reaction kinetics and other properties of RPUFs.

1.1.3 General Properties and Application

Rigid polyurethane foams (RPUFs) are widely utilized in various industries due to their remarkable mechanical properties, such as compressive strength, tensile strength, and dimensional stability. These properties are significantly influenced by the foam's density and cellular structure, which ultimately determine its suitability for different applications.

Compressive Strength. The compressive strength of RPUFs is strongly related to their density, which is mainly affected by the polyol and isocyanate formulas used in the synthesis process. High-density foams have a dense cellular structure, which allows for more material mass per unit area and increases the foam's modulus and strength. Studies indicate a consistent correlation between higher foam density and increased compressive strength [31–34].

The density of RPUF is affected by numerous factors, such as the types and concentrations of blowing agents and additives. Adding different fillers can significantly alter the density and, consequently, the mechanical properties of the foam. Adjusting the density of RPUFS enables the normalization of compressive strength, which aids in accurately assessing the material's inherent strength [35, 36]. This normalization process is essential because it provides a consistent method for evaluating the mechanical properties of RPUFs with different densities.

The polyol-to-isocyanate ratio variability significantly affects crosslinking density, subsequently influencing the mechanical properties of the foam. An increase in polyol content generally results in foam that has improved compressive strength due to the increased stiffness of the polymer matrix. However, if the ratio of polyol to isocyanate is imbalanced—specifically, if the polyol content is higher than that of isocyanate—it can lead to incomplete polymerization. This imbalance produces foams with reduced compressive strength [37, 38].

Tensile Properties. The tensile properties of RPUFs, including tensile strength and elongation at break, are essential for determining their operational integrity under tensile loads. The properties of RPUFs affect their ability to withstand stress before they fail, as well as the extent to which they can stretch before breaking. Understanding these mechanical properties is crucial, particularly in industries like automotive and construction, where RPUFs are subject to continuous mechanical stresses.

Tensile strength, the maximum stress that RPUFs can withstand during elongation, is closely associated with the foam structure's macroscopic and microscopic configurations. The connectivity among cell struts within the foam significantly affects stress distribution throughout the material on a macroscopic level. An interconnected cell network can enhance tensile strength by promoting uniform stress distribution, thus reducing the likelihood of localized structural failures [39]. Suboptimal strut connectivity can create localized weaknesses, reducing tensile strength and early material failure.

The uniformity of cell size distribution in RPUFs significantly affects their tensile properties at the microscopic level. Foams with uniform cell sizes generally exhibit improved mechanical properties, such as higher tensile strength and greater elongation at break. This uniformity ensures consistent stress distribution throughout the material, which reduces the risk of structural weaknesses that could lead to failure [39, 40]. In contrast, foams with varying cell sizes may experience uneven stress distribution, potentially compromising their tensile strength and elongation properties.

Elongation at break measures the degree of stretchability of the material before failure, influenced by the structural properties of the foam. RPUFs generally demonstrate reduced elongation rates relative to more flexible materials; nonetheless, formulation variations, including adding particular fillers or additives, can improve this characteristic. These modifications promote ductile behavior in the foam, enhancing its elongation capacity [41]. The crosslinking density within the polymer matrix, influenced by the polyol-to-isocyanate ratio, is crucial for regulating tensile strength and elongation at break. Higher crosslink density generally enhances tensile strength but can reduce elongation at break, resulting in decreased material flexibility [42].

Dimensional Stability. Dimensional stability is a critical characteristic of RPUFs, vital in applications requiring precise dimensions, such as insulating panels and aircraft components. This attribute pertains to the material's capacity to maintain its form and dimensions under diverse environmental conditions, such as temperature fluctuations and mechanical stresses. Dimensional stability is critical for adequately operating RPUFs in specific applications, as any fluctuation in dimensions can jeopardize insulating performance and structural integrity.

Higher-density RPUFs provide more excellent dimensional stability than lower-density equivalents. The improvement is due to the denser arrangement of polymer chains and cell struts in high-density foams, which increases resistance to deformation under thermal and mechanical loads and is critical for applications requiring exact dimensional stability [43]. High-density RPUFs generally exhibit a reduced coefficient of thermal expansion, thereby mitigating the extent of expansion or contraction resulting from temperature variations. This characteristic is especially vital in the building industry for insulating panels, as even little dimensional alterations can result in gaps, substantially affecting thermal efficiency [44].

The crosslinking density of RPUFs substantially affects their structural response to thermal stresses. Higher crosslinking density increases the rigidity and deformation resistance of the foam, resulting in more excellent dimensional stability. This component is critical in high-temperature situations because the material must retain its geometric and insulating properties [45]. Industrial standards require RPUFs in insulating panels to have a linear dimensional change of no more than 3% at temperatures near 70 °C, emphasizing the significance of dimensional stability in functional performance [44].

The microscopic structure of RPUFs, namely the distribution and connection of cell sizes, is critical in determining dimensional stability. The homogeneous distribution of cell sizes results in a more uniform mechanical reaction across the foam, lowering the possibility of localized deformations [46]. Cellular architectural irregularities can lead to vulnerabilities that are more susceptible to significant dimensional changes when stressed [36]. Improving the formulation of RPUFs to achieve a homogeneous cell structure is critical for increasing their dimensional stability.

Industrial Applications. RPUFs are known for their mechanical qualities, particularly compressive strength and dimensional stability, which are critical for use in the construction and automotive industries. In construction, RPUFs are generally used as insulation materials, utilizing their low thermal conductivity and high compressive strength to improve energy efficiency. The compressive strength of RPUFs is directly related to their density and cellular structure, with larger densities often resulting in greater strength. Jasiūnas et al. found that higher apparent densities in RPUFs correspond with higher compressive strengths, reaching up to 254 kPa. Furthermore, certain additives can increase these strengths, which is critical for designing RPUFs that can tolerate high structural stresses while offering efficient thermal insulation [47].

In the construction industry, RPUFs are used to fill wall cavities and roof gaps, resulting in more energy-efficient architectural designs. The improved insulative qualities of RPUFs can dramatically reduce energy expenditures related to heating and cooling, potentially saving up to 30% compared to standard insulating materials [48]. RPUFs are highly effective insulators, with thermal conductivity as low as 0.020 W/m·K [49].

In the automobile industry, RPUFs are valued for their lightweight nature and capacity to absorb energy, improving vehicle safety and performance. The compressive strength of these foams is critical since it directly affects the foam's ability to withstand impacts during collisions. RPUFs, enhanced with flame retardants and other additives, are increasingly used as structural components in automobiles [49, 50]. Furthermore, the lightweight properties of RPUFs contribute to overall vehicle weight reduction, increasing fuel efficiency and giving a significant benefit in satisfying tough emissions requirements [51].

The deformation and energy dispersion characteristics of RPUFs during crashes are critical for improving vehicle safety by dramatically reducing forces transferred to passengers. Studies have shown that RPUFs are effective at dispersing energy within crash structures and interior vehicle components [52]. By altering the density and cellular structure of RPUFs, manufacturers can adapt the foams' mechanical qualities to satisfy certain safety standards.

RPUFs' complete mechanical qualities, including compressive strength and dimensional stability, are critical across various applications in the construction and automotive industries. Their ability to provide exceptional thermal insulation while supporting structural loads makes them perfect for complicated, energy-efficient building designs. RPUFs are essential in the automotive industry because they lower weight and improve safety through their energy absorption characteristics. Ongoing research and development into the density and cellular architecture of RPUFs is expected to improve their functionality and effectiveness in these crucial sectors.

1.1.4 Thermal Insulation

Thermal insulation is essential for minimizing energy usage and enhancing energy efficiency in various applications, particularly in the refrigeration and building industries. RPUFs are highly valued in these sectors due to their low thermal conductivity, which makes them excellent insulators. This low conductivity is crucial in maintaining desired temperatures with minimal energy expenditure, thus significantly reducing operational costs.

Thermal Conductivity and Efficiency. RPUFs are renowned for their outstanding insulating capabilities, which are inextricably related to the distinct cellular structure of a network of closed cells. This structural arrangement is critical in drastically reducing heat transfer establishing RPUFs as highly efficient thermal insulators. The closed-cell design reduces convective heat transfer by restricting air movement within the foam, while the encapsulated gas within these cells, which has poor thermal conductivity, improves insulation efficacy [53, 54].

In practical applications, RPUFs are widely used in refrigeration systems and energy-efficient constructions. Their low thermal conductivity enables better temperature retention in refrigeration equipment, lowering energy usage and operational expenses. Comparative studies show that using RPUFs for building insulation can result in up to 30% energy cost savings, which is significantly better than conventional insulation materials, especially given the pressures of rising energy prices and stringent environmental regulations to reduce building energy usage [48].

The thermal performance of RPUFs is determined by various parameters, including the type of blowing agent used and the dimensions of the cellular structure. Blowing agents play an important part in the foaming process by creating gas-filled cells that are essential for insulation. Recent transitions have been observed from classic hydrochlorofluorocarbons (HCFCs) to more environmentally friendly alternatives such as hydrofluoroolefins (HFOs). HFOs, such as HFO-1336mzz-Z, have lower global warming potentials than HCFCs and have equivalent thermal insulation qualities while dramatically reducing environmental effects [55].

Adjusting the size of the foam cells can change the insulating properties of the material. Smaller cell sizes increase thermal resistance, thereby reducing channels for heat transmission. In contrast, larger cells may enhance the foam's density and structural integrity, but if not carefully managed, they could negatively impact thermal performance [56]. Additionally, the choice of blowing agents plays a significant role in determining thermal conductivity. For instance, cyclopentane, when used as a blowing agent, creates foams with lower thermal conductivity compared to those produced with water, due to its advantageous physical characteristics [57].

Aging and Thermal Performance. The thermal insulation properties of RPUFs are subject to degradation over time, influenced by factors such as gas diffusion from the foam and the physical deterioration of the foam structure. As RPUFs age, the gas encapsulated within their closed-cell structure may permeate out, resulting in increased thermal conductivity and diminished insulation effectiveness. This effect is particularly marked in foams with high initial gas concentrations, where gas diffusion through the polymer matrix substantially affects thermal performance [58, 59]. Moreover, environmental stressors such as moisture and temperature fluctuations can exacerbate the physical degradation of the foam, further eroding mechanical integrity and insulating properties [60].

Recent studies have concentrated on elucidating the aging effects on RPUF thermal efficiency. Research indicates that incorporating specific fillers and altering foam formulations may ameliorate these adverse impacts. The integration of natural fibers and other bio-based materials has been explored to bolster the durability and thermal stability of RPUFs. Notably, the addition of cellulose whiskers has proven to enhance mechanical properties and curtail gas diffusion rates, thus prolonging the foam's functional lifespan [61]. Additionally, optimizing cell structure via controlled foaming processes can yield more robust cell configurations that better withstand physical degradation, thereby preserving insulative properties over extended periods [62].

Advances in polymer chemistry have also played a pivotal role in mitigating the aging process of RPUFs. The development of high-functionality polyols, for example, has been shown to increase the crosslinking density within the polymer matrix, which not only enhances thermal stability but also diminishes the likelihood of gas diffusion [63]. Furthermore, the integration of nanoparticles and other reinforcing agents has demonstrated significant potential in augmenting the barrier properties of the foam. The employment of expandable graphite and other nanofillers, for instance, can effectively reduce gas diffusion rates by creating a more intricate path for gas molecules, thereby enhancing the overall thermal insulation performance [64–66]. These modifications serve dual purposes: they not only improve thermal properties but also contribute to enhanced fire retardancy, a critical attribute for many RPUF applications [67].

Understanding the benefits and limitations of RPUFs in thermal insulation is essential for optimizing their application in critical areas such as refrigeration and energy-efficient building construction. Ongoing research and development are vital for enhancing the effectiveness and lifespan of these materials. By continuously improving formulations and gaining insights into the aging processes and thermal performance, scientists and engineers can create RPUFs that meet the evolving demands for energy efficiency and environmental sustainability. This body of work not only deepens our understanding of material science but also paves the way for future innovations in thermal insulation technology.

1.1.5 Emerging Applications

Rigid polyurethane foams (RPUFs) are increasingly recognized for their versatility and potential in a variety of industries beyond their conventional roles. Known for their lightweight, adaptable, and superior insulative properties, RPUFs are paving the way for innovative applications across new sectors. These foams continue to evolve, bridging the gap between traditional uses and cutting-edge technologies, thereby contributing significantly to advancements in sustainability and advanced engineering.

Innovation. Rigid polyurethane foams (RPUFs) have become pivotal in advancing technology and sustainability across several key industries, including aerospace, electronics, and eco-friendly construction, due to their unique combination of low density, high thermal insulation, and robust mechanical strength.

Aerospace Applications. In the aerospace sector, RPUFs are integral as lightweight structural components, playing a crucial role in reducing vehicle weight, enhancing fuel efficiency, and decreasing greenhouse gas emissions. The low-density attribute of RPUFs facilitates the design of lighter aircraft, which inherently require less energy for operation, thus improving overall fuel economy [40]. RPUFs are utilized in various aerospace components, such as interior cabin panels and insulation materials, which must endure extreme environmental conditions including high temperatures and pressures. Their durability and environmental resistance render them highly suitable for these purposes [68]. A prominent example is the incorporation of RPUFs in the Boeing 787 Dreamliner's insulation systems, where they contribute significantly to weight reduction and enhanced thermal management [69].

Electronics. In the electronics industry, RPUFs are vital in thermal management systems, where they play a key role in dissipating heat to augment performance and extend the lifespan of high-performance electronics such as computers and servers. The closed-cell structure of RPUFs offers excellent thermal insulation, crucial for maintaining optimal device operating temperatures. This property is especially significant in data centers, where preventing overheating is essential to avoid equipment failure and reduce cooling-related energy consumption [70]. Due to its unique cell structure, integrating RPUFs into server racks could potentially decrease thermal conductivity, thereby enhancing airflow and cooling efficiency [32]. Additionally, RPUFs can be tailored to possess specific thermal conductivity properties, to meet the exacting demands of contemporary electronic applications [71].

Eco-friendly Construction. Within the domain of eco-friendly construction, RPUFs are celebrated for their insulating capabilities in green buildings. Their low thermal conductivity helps maintain comfortable indoor environments while reducing the energy required for heating and cooling [72]. Commonly, RPUFs are applied in external wall insulation, roofing systems, and as core materials in Structural Insulated Panels (SIPs), which are pivotal for attaining high environmental standards in modern building practices [73]. The development of RPUF formulations incorporating bio-based polyols sourced from renewable materials further bolsters their sustainability credentials. This is in line with the increasing demand for construction materials that are both environmentally friendly and effective in performance. For instance, RPUFs enhanced with natural fillers like lignin or rapeseed oil-based polyols demonstrate promising outcomes in reducing the carbon footprint of building materials while sustaining superior thermal insulation characteristics [74].

1.1.6 Green Technology

Rigid polyurethane foams (RPUFs) are instrumental in promoting green technology, especially in applications such as building insulation where sustainability is a focus. Their inherent properties—low thermal conductivity, high compressive strength, and lightweight nature—render them particularly suitable for enhancing thermal efficiency in buildings, thus reducing energy consumption and overall carbon footprints.

Energy Conservation. RPUFs are renowned for their superior thermal insulation capabilities, which play a critical role in decreasing the energy needed for heating and cooling buildings. With thermal conductivities as low as 0.020 W/m·K, RPUFs provide effective temperature control within structures, reducing the dependency on external heating or cooling systems [33, 75, 76]. This improved efficiency not only lowers energy costs for occupants but also contributes to diminishing greenhouse gas emissions from energy production. For instance, buildings insulated with RPUFs have been shown to realize energy savings of up to 30% compared to those using traditional insulation materials, as mentioned in previous sections [48].

Additionally, the use of RPUFs in various structural components—walls, roofs, and floors—significantly enhances the overall energy performance of buildings. Their incorporation into Structural Insulated Panels (SIPs) and insulation for HVAC systems further optimizes this efficiency, positioning them as a preferred choice in eco-friendly construction practices [73].

Sustainability. Beyond their role in insulation, RPUFs contribute to sustainable industrial practices through the integration of bio-based polyols in their production. Traditionally, RPUFs are derived from petroleum-based polyols, which pose environmental degradation risks and contribute to greenhouse gas emissions. However, recent advances have shifted focus toward synthesizing RPUFs using bio-based polyols from renewable sources, such as vegetable oils and recycled polyurethane waste, which significantly lessen environmental impacts [77–79]. For instance, palm oil-based polyols have shown potential in producing RPUFs that retain the mechanical and thermal properties of their petroleum-derived counterparts [80, 81].

Incorporating bio-based materials not only enhances the sustainability profile of RPUFs but also reduces their lifecycle environmental impact. Utilizing renewable resources allows manufacturers to decrease reliance on fossil fuels and reduce the carbon footprint associated with foam production [78, 82]. Additionally, the recycling of post-industrial polyurethane foam waste into new RPUF formulations exemplifies the principles of a circular economy, further advancing sustainability within the industry [77].

The future of RPUFs in innovative and green technology fields looks promising. Ongoing research aims to enhance their environmental benefits and performance in advanced applications. Studies predict that foam technology will continue to evolve, driven by the demand for materials that provide both high performance and reduced environmental impact. RPUFs are set to play a crucial role in this transformation, signaling a significant shift in how materials science contributes to sustainable and advanced engineering solutions.

1.2 Introduction to RPUF Computational Modeling

1.2.1 Computational Modeling

In material science, computational modeling is a transformative tool that enables researchers to predict and optimize complex materials' behavior and performance. Within the context of rigid polyurethane foams (RPUFs), computational modeling acts as a virtual testing ground, allowing scientists to analyze and refine formulations before physical prototyping. Given the intricate chemistry of RPUF production—characterized by the exothermic reaction between polyols and isocyanates and influenced by the presence of catalysts and blowing agents—computational modeling is essential for achieving precise control over foam properties.

Thermodynamic Models. Thermodynamic modeling examines the heat and energy exchanges within RPUF systems, which is particularly relevant due to the exothermic nature of the foaming reaction. Thermodynamic models simulate energy transformations as the foam expands, helping researchers to predict thermal conductivity and stability. For example, during the formation of RPUFs, heat generated from the reaction can impact the integrity of the cell structure. Thermodynamic simulations allow researchers to model the influence of different ratios of reactants on temperature profiles, as demonstrated by Zhao et al. [83]. Their study modeled the thermal effects of varying polyol-to-isocyanate ratios, providing critical insights into how adjustments in the formulation can manage the rate of heat release, thereby minimizing issues related to foam collapse and thermal degradation.

Thermodynamic models are essential for applications where insulation is a priority, such as in refrigeration or building construction. By optimizing thermal profiles through simulation, researchers can design RPUFs with superior insulation properties that maintain stability even under extreme temperature variations. As energy-efficient buildings and devices gain prominence, thermodynamic models will continue to be vital for advancing RPUF applications in insulation technology.

Kinetic Models. Kinetic models focus on reaction rates within RPUF formulations, providing a detailed view of how catalysts, blowing agents, and other additives interact to influence foam expansion and density. These models simulate the speed and progression of chemical reactions, which affect factors such as cell size and distribution. Kinetic modeling is especially valuable in designing foams for applications where density and mechanical strength are critical. For instance, Ghoreishi et al. [84] applied kinetic models to evaluate how different hydroxyl group structures in polyols impact reaction speed, enabling them to identify optimal polyol-catalyst combinations that improve foam durability and load-bearing properties.

The impact of kinetic modeling extends beyond simply optimizing density; it enables a detailed understanding of how to achieve desired foam structures across a range of environmental conditions. Researchers can adjust parameters like catalyst concentration and reactant ratios in simulations to explore new formulations that balance the speed of reaction with structural integrity. These insights are crucial for producing RPUFs with specific properties tailored to their intended applications, such as lightweight automotive parts or structural elements in construction.

Mechanical Models. Mechanical models evaluate RPUF's physical responses to stress and strain, a key consideration in industries where foams are used as structural components. Finite element modeling (FEM) is commonly applied to simulate the distribution of forces within foam structures, enabling researchers to predict the foam's performance under various loading conditions. Whisler and Kim [85] used FEM to study how RPUFs respond to compressive and tensile forces, demonstrating that specific formulations could be adjusted to meet the requirements of applications that demand high mechanical resilience, such as crash-resistant automotive interiors.

Mechanical models allow for the exploration of RPUFs' structural limits and durability. In aerospace and automotive applications, where materials are exposed to both dynamic and static loads, understanding the mechanical performance of RPUFs under simulated conditions is critical. These models allow engineers to optimize foam density and cell structure to achieve maximum durability without compromising weight—a priority in industries focused on lightweight design for fuel efficiency.

Together, these computational models enable researchers to refine RPUF formulations with a high degree of precision. By simulating material behavior under various environmental conditions—including temperature, pressure, and mechanical stress—these models allow scientists to reduce the dependency on trial-and-error experimentation. Computational predictions provide a more efficient route to achieving materials with customized properties, contributing to innovations across multiple industries that rely on RPUFs.

1.2.2   Brief History and Timeline

The development of computational modeling in RPUF research has paralleled advancements in both materials science and computational technology. Initially, RPUF properties were determined through empirical experimentation, which relied heavily on physical testing and manual adjustments to formulations. Early models were simplistic, based on linear approximations of variable relationships that offered limited insights into the complex, nonlinear behavior observed in foam systems.

Finite Element Analysis (FEA) marked a significant advancement in RPUF modeling during the late twentieth century. With FEA, researchers gained the ability to simulate mechanical stress and strain distribution within foam structures, which was especially useful for understanding material behavior in applications where structural integrity was crucial. FEA became a standard method for assessing foam performance in crash safety applications in the automotive sector, where RPUFs are valued for their combination of strength and lightweight properties. The work of Ghoreishi et al. [84] illustrates how FEA enabled the industry to develop more robust RPUFs that meet stringent safety standards.

Over the past two decades, the integration of machine learning (ML) and artificial intelligence (AI) has revolutionized computational modeling in RPUF research. ML algorithms analyze large datasets to reveal complex relationships among formulation parameters and material properties, allowing for the development of highly predictive models. For instance, Alfeche et al. [18] applied ML algorithms to explore optimal combinations of bio-based polyols, substantially reducing the need for experimental validation by providing predictive insights that guided the formulation process.

The timeline of RPUF computational modeling demonstrates the increasing precision and sophistication of modeling techniques. From linear empirical models to multiscale simulations incorporating AI-driven predictions, the field has evolved to accommodate the growing demand for highly specialized RPUFs with tailored properties. As computational capacity continues to expand, the next generation of models will likely incorporate real-time predictive capabilities, integrating aspects of thermodynamic, kinetic, and mechanical behavior into a single, cohesive framework.

1.2.3 Impact on Academia and Industry

Academic Relevance. In academia, computational modeling has transformed RPUF research by reducing the need for extensive physical testing and enabling a deeper understanding of the material's behavior under various conditions. Computational models allow researchers to explore hypotheses and test variables in a controlled, virtual environment, where formulations can be refined and optimized before physical synthesis. Al-Moameri et al. [86] demonstrated how kinetic and thermodynamic models can support the development of bio-based polyols for sustainable RPUFs, providing insights that contribute to the advancement of eco-friendly materials without the need for extensive laboratory resources.

Computational modeling also facilitates academic research into the molecular interactions that influence macroscopic properties in RPUFs. For instance, thermodynamic models that simulate the behavior of physical blowing agents (PBAs) and catalysts within RPUFs have illuminated mechanisms that govern thermal conductivity and insulation properties. This fundamental research supports a broad range of applications and contributes to the academic community's knowledge base, laying the groundwork for future innovations in both RPUFs and other polymeric materials.

Industrial Relevance. Computational modeling has transformed RPUF manufacturing by enabling predictive optimization of formulations and production processes. In industries such as automotive and construction, where RPUFs are integral to the production of lightweight and thermally insulating components, computational models help manufacturers streamline product development by reducing material waste and improving consistency. Dingcong et al. [87] illustrated that computational models allowed for faster formulation testing in the automotive sector, where foams must meet high standards for impact resistance and structural integrity. Case studies highlight the impact of computational modeling across various industrial sectors. In the automotive industry, for example, RPUFs are often used for energy-absorbing structures in vehicle interiors. Mechanical simulations predict how these foams will respond to forces in a crash scenario, allowing engineers to refine foam density and cell distribution for maximum safety. In the construction industry, RPUFs serve as insulative materials. Computational models help manufacturers achieve optimal thermal conductivity by adjusting PBA concentrations and refining the cellular structure. In the marine industry, where RPUFs are used for buoyancy and impact resistance, simulations enable the design of formulations that perform reliably in extreme conditions.

1.2.4 Advancements in RPUF Computational Modeling

Recent Developments. Recent advancements in computational modeling have seen the integration of AI and machine learning, which enhance the precision and predictive capacity of traditional models. Machine learning algorithms excel at handling the nonlinear, multivariable relationships that characterize RPUF formulations. Alfeche et al. [18] used machine learning to automate the optimization of bio-based RPUF formulations, achieving material properties that aligned with specific application requirements while reducing experimental workload.

The incorporation of multiscale modeling techniques has enabled researchers to study both microscopic and macroscopic behaviors of RPUFs within a single framework. For example, kinetic models simulate the interactions at the molecular level, while mechanical simulations predict performance at the structural level. This convergence of modeling techniques allows researchers to understand the impact of small-scale chemical changes on large-scale mechanical properties, enhancing the customization of RPUFs for high-performance applications.

1.2.5 Future Prospects

The future of computational modeling in RPUF research lies in technologies such as quantum computing and big data analytics. Quantum computing could enable real-time molecular simulations, allowing researchers to examine interactions between individual atoms within the polymer matrix. This level of detail could provide insights into how specific molecular arrangements impact properties like flexibility, strength, and thermal stability, paving the way for RPUFs with unprecedented performance characteristics.

Big data analytics, combined with machine learning, could lead to a fully automated approach to RPUF formulation. By analyzing historical data from past experiments, big data algorithms could identify optimal formulation parameters for new applications, effectively automating the process of material design. This shift toward data-driven, automated modeling represents a significant advancement in RPUF research, with implications for industries seeking highly specialized materials that can be developed faster and more efficiently.

The continued advancement of computational modeling promises to push the boundaries of material science, allowing researchers and industry professionals to develop RPUFs that meet increasingly specific and demanding requirements. As these tools evolve, they will provide unprecedented insights into the fundamental properties of RPUFs, reinforcing their role as essential materials in modern industry.

References

1. A. Das, P. Mahanwar, A brief discussion on advances in polyurethane applications. Adv. Ind. Eng. Polym. Res. **3**(3) (2020). https://doi.org/10.1016/j.aiepr.2020.07.002. Available: https://www.sciencedirect.com/science/article/pii/S2542504820300269
2. M. Szycher, *Szycher's Handbook of Polyurethanes* (Crc Press, Boca Raton, Fl, 2013)
3. D. Eaves, *Handbook of Polymer Foams* (Rapra Technology, Shawbury, 2004)
4. R. Koschade, *Sandwich Construction Method* (Wiley-Vch, Weinheim, Cambridge, 2001)
5. H Grünbauer et al., *Rigid Polyurethane Foams* (CRC Press eBooks, 2004), https://doi.org/10.1201/9780203506141.ch7
6. A.M. Borrero-López, V. Nicolas, Z. Marie, A. Celzard, V. Fierro, A review of rigid polymeric cellular foams and their greener tannin-based alternatives. Polymers **14**(19), 3974 (2022). https://doi.org/10.3390/polym14193974
7. C. Defonseka, *Flexible Polyurethane Foams* (Walter de Gruyter GmbH & Co KG, 2019)
8. F. Zafar, Eram Sharmin, *Polyurethane/Monograph* (Intech, Rijeka, Croatia, 2012)
9. M. Barikani, N. Fazeli, M. Barikani, Study on thermal properties of polyurethane-urea elastomers prepared with different dianiline chain extenders. J. Polym. Eng. **33**(1), 87–94 (2013). https://doi.org/10.1515/polyeng-2012-0137
10. H. Kalita, N. Karak, Biobased hyperbranched shape-memory polyurethanes: effect of different vegetable oils. J. Appl. Polym. Sci. **131**(1) (2013). https://doi.org/10.1002/app.39579
11. C. Zhang, J. Hu, X. Li, Y. Wu, J. Han, Hydrogen-Bonding interactions in hard segments of shape memory polyurethane: toluene diisocyanates and 1,6-hexamethylene diisocyanate. a theoretical and comparative study. J. Phys. Chem. A **118**(51), 12241–12255 (2014). https://doi.org/10.1021/jp508817v
12. D.K. Chattopadhyay, D.C. Webster, Thermal stability and flame retardancy of polyurethanes. Prog. Polym. Sci. **34**(10), 1068–1133 (2009). https://doi.org/10.1016/j.progpolymsci.2009.06.002
13. S. Kaewunruen, C. Ngamkhanong, M. Papaelias, C.J. Roberts, Wet/dry influence on behaviors of closed-cell polymeric cross-linked foams under static, dynamic and impact loads. Constr. Building Mater. **187**, 1092–1102 (2018). https://doi.org/10.1016/j.conbuildmat.2018.08.052
14. A.N. Gent, K.C. Rusch, Permeability of open-cell foamed materials. J. Cell. Plast. **2**(1), 46–51 (1966). https://doi.org/10.1177/0021955x6600200112
15. C.J. Omisol et al., Flexible polyurethane foams modified with novel coconut monoglycerides-based polyester polyols. ACS Omega (2024). https://doi.org/10.1021/acsomega.3c07312
16. J. Peyrton, L. Avérous, Structure-properties relationships of cellular materials from biobased polyurethane foams **145**, 100608–100608 (2021). https://doi.org/10.1016/j.mser.2021.100608
17. accufoam, Structural spray foam strength [Expert Guide] | Accufoam®. Accufoam (2021). Available: https://accufoam.com/spray-foam-structural-strength/. Accessed 10 Nov 2024

18. L. Adeliene et al., In silico investigation of the impact of reaction kinetics on the physico-mechanical properties of coconut-oil-based rigid polyurethane foam. Sustainability **15**(9), 7148–7148 (2023). https://doi.org/10.3390/su15097148

19. C. Li, J. Han, Q. Huang, H. Xu, J. Tao, X. Li, Microstructure development of thermoplastic polyurethanes under compression: the influence from first-order structure to aggregation structure and a structural optimization. Polymer **53**(5), 1138–1147 (2012). https://doi.org/10.1016/j.polymer.2012.01.019

20. H.J. Qi, M.C. Boyce, Stress–strain behavior of thermoplastic polyurethanes. Mech. Mater. **37**(8), 817–839 (2005). https://doi.org/10.1016/j.mechmat.2004.08.001

21. P. Krol, Synthesis methods, chemical structures and phase structures of linear polyurethanes. Properties and applications of linear polyurethanes in polyurethane elastomers, copolymers and ionomers. Prog. Mater. Sci. **52**(6), 915–1015 (2007). https://doi.org/10.1016/j.pmatsci.2006.11.001

22. M. Ionescu, *Mihail Ionescu: Polyols for Polyurethanes*, vol 2 (Walter de Gruyter GmbH & Co KG, 2019)

23. A. Lapprand, F. Boisson, F. Delolme, F. Méchin, J.-P. Pascault, Reactivity of isocyanates with urethanes: conditions for allophanate formation. Polym. Degrad. Stab. **90**(2), 363–373 (2005). https://doi.org/10.1016/j.polymdegradstab.2005.01.045

24. S. Hee Kim, H. Lim, J. Chul Song, B. Kyu Kim, Effect of blowing agent type in rigid polyurethane foam. J. Macromol. Sci., Part A **45**(4), 323–327 (2008). https://doi.org/10.1080/10601320701864260

25. K.H. Choe, D.S. Lee, W.J. Seo, W.N. Kim, Properties of rigid polyurethane foams with blowing agents and catalysts. Polym. J. **36**(5), 368–373 (2004). https://doi.org/10.1295/polymj.36.368

26. UN Environment, Foam. Ozonaction (2018). Available: https://www.unep.org/ozonaction/what-we-do/foam. Accessed 10 Nov 2024

27. J.C. Bondaug et al., Development of a catalyst system for enhanced properties of coconut diethanolamide-based rigid poly(urethane-urea) foam. ACS Appl. Polym. Mater. **6**(11), 6875–6887 (2024). https://doi.org/10.1021/acsapm.4c01187

28. H. Sardon, A. Pascual, D. Mecerreyes, D. Taton, H. Cramail, J.L. Hedrick, Synthesis of polyurethanes using organocatalysis: a perspective. Macromolecules **48**(10), 3153–3165 (2015). https://doi.org/10.1021/acs.macromol.5b00384

29. D.C. Fondots, Developments in amine catalysts for urethane foam. J. Cell. Plast. **11**(5), 250–255 (1975). https://doi.org/10.1177/0021955x7501100503

30. K.C. Frisch, History of science and technology of polymeric foams. J. Macromol. Sci.: Part A Chem. **15**(6), 1089–1112 (1981). https://doi.org/10.1080/00222338108066455

31. A. Strąkowska, S. Członka, K. Strzelec, POSS compounds as modifiers for rigid polyurethane foams (composites). Polymers (Basel) **11**(7), 1092 (2019). https://doi.org/10.3390/polym11071092

32. M. Günther, A. Lorenzetti, B. Schartel, Fire phenomena of rigid polyurethane foams. Polymers (Basel) **10**(10), 1166 (2018). https://doi.org/10.3390/polym10101166

33. E. Akdogan, M. Erdem, M.E. Ureyen, M. Kaya, Rigid polyurethane foams with halogen-free flame retardants: thermal insulation, mechanical, and flame retardant properties. J. Appl. Polym. Sci. **137**(1) (2020). https://doi.org/10.1002/app.47611

34. D. Dukarska, J. Walkiewicz, A. Derkowski, R. Mirski, Properties of rigid polyurethane foam filled with sawdust from primary wood processing. Materials (Basel) **15**(15), 5361 (2022). https://doi.org/10.3390/ma15155361

35. S. Shin, H. Kim, J. Liang, S. Lee, D. Lee, Sustainable rigid polyurethane foams based on recycled polyols from chemical recycling of waste polyurethane foams. J. Appl. Polym. Sci. **136**(35) (2019). https://doi.org/10.1002/app.47916

36. S. Wang, F. Yang, W. Sun, X. Xu, Y. Deng, Bio-based flame-retardant rigid polyurethane foam with high content of soybean oil polyols containing phosphorus. J. Am. Oil Chem. Soc. **100**(7), 561–577 (2023). https://doi.org/10.1002/aocs.12706

37. Y. Luo, Z. Ye, J. Yan, Effects of high functionality polyol on the structure and properties of rigid polyurethane foam. Plast. Rubber Compos. **51**(3), 154–161 (2022). https://doi.org/10.1080/14658011.2021.1959132

38. S.H. Kim et al., Nanoclay reinforced rigid polyurethane foams. J. Appl. Polym. Sci. **117**(4), 1992–1997 (2010). https://doi.org/10.1002/app.32116
39. M. Ridha, V.P.W. Shim, Microstructure and tensile mechanical properties of anisotropic rigid polyurethane foam. Exp. Mech. **48**(6), 763–776 (2008). https://doi.org/10.1007/s11340-008-9146-0
40. M. Leszczyńska et al., Vegetable fillers and rapeseed oil-based polyol as natural raw materials for the production of rigid polyurethane foams. Materials (Basel) **14**(7), 1772 (2021). https://doi.org/10.3390/ma14071772
41. A. Kairytė, S. Członka, R. Boris, S. Vėjelis, Vacuum-Based impregnation of liquid glass into sunflower press cake particles and their use in bio-based rigid polyurethane foam. Materials (Basel) **14**(18), 5351 (2021). https://doi.org/10.3390/ma14185351
42. C.S. Lee, T.L. Ooi, C.H. Chuah, S. Ahmad, Rigid polyurethane foam production from palm oil-based epoxidized diethanolamides. J. Am. Oil Chem. Soc. **84**(12), 1161–1167 (2007). https://doi.org/10.1007/s11746-007-1150-5
43. R. de A. Delucis, W.L.E. Magalhães, C.L. Petzhold, S.C. Amico, Thermal and combustion features of rigid polyurethane biofoams filled with four forest-based wastes. Polym. Compos. **39**(S3) (2018). https://doi.org/10.1002/pc.24784
44. S. Członka, A. Strąkowska, K. Strzelec, A. Adamus-Włodarczyk, A. Kairytė, S. Vaitkus, Composites of rigid polyurethane foams reinforced with POSS. Polymers (Basel) **11**(2), 336 (2019). https://doi.org/10.3390/polym11020336
45. H. Yang et al., Aluminum hypophosphite in combination with expandable graphite as a novel flame retardant system for rigid polyurethane foams. Polym. Adv. Technol. **25**(9), 1034–1043 (2014). https://doi.org/10.1002/pat.3348
46. Y. Wang, K. Cui, B. Fang, F. Wang, Cost-effective fabrication of modified palygorskite-reinforced rigid polyurethane foam nanocomposites. Nanomaterials **12**(4), 609 (2022). https://doi.org/10.3390/nano12040609
47. L. Jasiūnas, S.T. McKenna, D. Bridžiuvienė, L. Miknius, Mechanical, thermal properties and stability of rigid polyurethane foams produced with crude-glycerol derived biomass biopolyols. J. Polym. Environ. **28**(5), 1378–1389 (2020). https://doi.org/10.1007/s10924-020-01686-y
48. L. Gao, G. Zheng, Y. Zhou, L. Hu, G. Feng, Improved mechanical property, thermal performance, flame retardancy and fire behavior of lignin-based rigid polyurethane foam nanocomposite. J. Therm. Anal. Calorim. **120**(2), 1311–1325 (2015). https://doi.org/10.1007/s10973-015-4434-2
49. E. Akdogan, M. Erdem, Improvement in physico-mechanical and structural properties of rigid polyurethane foam composites by the addition of sugar beet pulp as a reactive filler. J. Polym. Res. **28**(3), 80 (2021). https://doi.org/10.1007/s10965-021-02445-w
50. E.F. Kerche, L.K. Lazzari, B.F. de Bortoli, R.D. de Oliveira Polkowski, R.F.C. de Albuquerque, A systematic review of enhanced polyurethane foam composites modified with graphene for automotive industry. Graphene 2D Mater. **9**(1–2), 27–46 (2024). https://doi.org/10.1007/s41127-024-00073-x
51. A.L. Robinson, A.I. Taub, G.A. Keoleian, Fuel efficiency drives the auto industry to reduce vehicle weight. MRS Bull. **44**(12), 920–923 (2019). https://doi.org/10.1557/mrs.2019.298
52. H. Liang, Y. Zhao, S. Chen, F. Ma, D. Wang, Review of crashworthiness studies on cellular structures. Automot. Innov. **6**(3), 379–403 (2023). https://doi.org/10.1007/s42154-023-00237-0
53. J.C. Grumo, L.J.Y. Jabber, A.A. Lubguban, R.Y. Capangpangan, A.C. Alguno, Synthesis and characterization of bio-based rigid polyurethane foams with varying amount of blowing agent. Key Eng. Mater. **803**, 346–350 (2019). https://doi.org/10.4028/www.scientific.net/KEM.803.346
54. H. Wang, Y. Liu, L. Lin, Behavior characteristics and thermal energy absorption mechanism of physical blowing agents in polyurethane foaming process. Polymers (Basel) **15**(10), 2285 (2023). https://doi.org/10.3390/polym15102285
55. S. Wang et al., Experimental evaluation on low global warming potential HFO-1336mzz-Z as an alternative to HCFC-123 and HFC-245fa. J. Therm. Sci. Eng. Appl. **11**(3) (2019). https://doi.org/10.1115/1.4041881

56. C. Carriço, T. Fraga, V. Carvalho, V. Pasa, Polyurethane foams for thermal insulation uses produced from castor oil and crude glycerol biopolyols. Molecules **22**(7), 1091 (2017). https://doi.org/10.3390/molecules22071091

57. R. Boonachathong, B. Kaewnok, H. Widjaja, S. Amornraksa, Development of rigid polyurethane foam (RPUF) for imitation wood blown by distilled water and cyclopentane (CP). MATEC Web Conf. **187**, 02001 (2018). https://doi.org/10.1051/matecconf/201818702001

58. B. Aydoğan, N. Usta, Effects of dolomite and intumescent flame retardant additions on thermal and combustion behaviors of rigid polyurethane foams. J. Appl. Polym. Sci. **140**(16) (2023). https://doi.org/10.1002/app.53739

59. A. Acosta et al., Wood flour modified by poly(furfuryl alcohol) as a filler in rigid polyurethane foams: effect on water uptake. Polymers (Basel) **14**(24), 5510 (2022). https://doi.org/10.3390/polym14245510

60. E.F. Kerche, D.N. Bock, R. de Avila Delucis, W.L.E. Magalhães, S.C. Amico, Micro fibrillated cellulose reinforced bio-based rigid high-density polyurethane foams. Cellulose **28**(7), 4313–4326 (2021). https://doi.org/10.1007/s10570-021-03801-1

61. F. Luo, K. Wu, H. Guo, Q. Zhao, L. Liang, M. Lu, Effect of cellulose whisker and ammonium polyphosphate on thermal properties and flammability performance of rigid polyurethane foam. J. Therm. Anal. Calorim. **122**(2), 717–723 (2015). https://doi.org/10.1007/s10973-015-4766-y

62. I. Aranberri, S. Montes, E. Wesołowska, A. Rekondo, K. Wrześniewska-Tosik, H.-J. Grande, Improved thermal insulating properties of renewable polyol based polyurethane foams reinforced with chicken feathers. Polymers (Basel) **11**(12), 2002 (2019). https://doi.org/10.3390/polym11122002

63. Y. Luo, K. Pu, J. Gao, Y. Zhou, J. Wan, X. Bai, Thermal degradation behavior and kinetics of porous polymer based on high functionality components. J. Appl. Polym. Sci. **141**(18) (2024). https://doi.org/10.1002/app.55304

64. J. Cheng, W. Qu, S. Sun, Mechanical properties improvement and fire hazard reduction of expandable graphite microencapsulated in rigid polyurethane foams. Polym. Compos. **40**(S2) (2019). https://doi.org/10.1002/pc.24786

65. Y. Wang et al., Expandable graphite encapsulated by magnesium hydroxide nanosheets as an intumescent flame retardant for rigid polyurethane foams. J. Appl. Polym. Sci. **135**(39) (2018). https://doi.org/10.1002/app.46749

66. X. Zhang et al., A facile strategy to fabricate microencapsulated expandable graphite as a flame-retardant for rigid polyurethane foams. J. Appl. Polym. Sci. **132**(31) (2015). https://doi.org/10.1002/app.42364

67. A. Muhammed Raji, H.U. Hambali, Z.I. Khan, Z. Binti Mohamad, H. Azman, R. Ogabi, Emerging trends in flame retardancy of rigid polyurethane foam and its composites: a review. J. Cell. Plast. **59**(1), 65–122 (2023). https://doi.org/10.1177/0021955X221144564

68. X. Wang, H. Peng, T. Li, C. Lou, Y. Wang, J. Lin, Preparation and property evaluations of zeolite rigid foam composites. Polym. Compos. **40**(11), 4175–4185 (2019). https://doi.org/10.1002/pc.25278

69. S. Zhou et al., Weathering of roofing insulation materials under multi-field coupling conditions. Materials (Basel) **12**(20), 3348 (2019). https://doi.org/10.3390/ma12203348

70. Z. Zhang et al. Long-Service-Life rigid polyurethane foam fillings for spent fuel transportation casks. (2023). https://doi.org/10.20944/preprints202312.1599.v1

71. K. Uram, M. Kurańska, J. Andrzejewski, A. Prociak, Rigid polyurethane foams modified with biochar. Materials (Basel) **14**(19), 5616 (2021). https://doi.org/10.3390/ma14195616

72. X. Gu, X. Wang, X. Guo, S. Liu, Q. Li, Y. Liu, Study and characterization of regenerated hard foam prepared by polyol hydrolysis of waste polyurethane. Polymers (Basel) **15**(6), 1445 (2023). https://doi.org/10.3390/polym15061445

73. H.M.C.C. Somarathna, S.N. Raman, D. Mohotti, A.A. Mutalib, K.H. Badri, The use of polyurethane for structural and infrastructure engineering applications: a state-of-the-art review. Constr. Build. Mater. **190**, 995–1014 (2018). https://doi.org/10.1016/j.conbuildmat.2018.09.166

74. M. Kurańska, J.A. Pinto, K. Salach, M.F. Barreiro, A. Prociak, Synthesis of thermal insulating polyurethane foams from lignin and rapeseed based polyols: a comparative study. Ind. Crops Prod. **143**, 111882 (2020). https://doi.org/10.1016/j.indcrop.2019.111882

75. H. Singh, Rigid polyurethane foam: a versatile energy efficient material. Key Eng. Mater. **678**, 88–98 (2016). https://doi.org/10.4028/www.scientific.net/KEM.678.88

76. Z. Wang, C. Wang, Y. Gao, Z. Li, Y. Shang, H. Li, Porous thermal insulation polyurethane foam materials. Polymers (Basel) **15**(18), 3818 (2023). https://doi.org/10.3390/polym15183818

77. I. Amundarain, R. Miguel-Fernández, A. Asueta, S. García-Fernández, S. Arnaiz, Synthesis of rigid polyurethane foams incorporating polyols from chemical recycling of post-industrial waste polyurethane foams. Polymers (Basel) **14**(6), 1157 (2022). https://doi.org/10.3390/polym14061157

78. M. Zieleniewska et al., Preparation and characterisation of rigid polyurethane foams using a rapeseed oil-based polyol. Ind. Crops Prod. **74**, 887–897 (2015). https://doi.org/10.1016/j.indcrop.2015.05.081

79. D. Abril-Milán, O. Valdés, Y. Mirabal-Gallardo, A.F. de la Torre, C. Bustamante, J. Contreras, Preparation of renewable bio-polyols from two species of colliguaja for rigid polyurethane foams. Materials (Basel) **11**(11), 2244 (2018). https://doi.org/10.3390/ma11112244

80. N. Sukhawipat, L. Saengdee, P. Pasetto, J. Junthip, E. Martwong, Sustainable rigid polyurethane foam from wasted palm oil and water hyacinth fiber composite—a green sound-absorbing material. Polymers (Basel) **14**(1), 201 (2022). https://doi.org/10.3390/polym14010201

81. X. Zhou, M.M. Sain, K. Oksman, Semi-rigid biopolyurethane foams based on palm-oil polyol and reinforced with cellulose nanocrystals. Compos. Part A Appl. Sci. Manuf. **83**, 56–62 (2016). https://doi.org/10.1016/j.compositesa.2015.06.008

82. K. Uram, A. Prociak, L. Vevere, R. Pomilovskis, U. Cabulis, M. Kirpluks, Natural oil-based rigid polyurethane foam thermal insulation applicable at cryogenic temperatures. Polymers (Basel) **13**(24), 4276 (2021). https://doi.org/10.3390/polym13244276

83. Y. Zhao, M.J. Gordon, A. Tekeei, Modeling reaction kinetics of rigid polyurethane foaming process. J. Appl. Polym. Sci. **130**, 1131–1138 (2013)

84. R. Ghoreishi, Y. Zhao, G.J. Suppes, Reaction modeling of urethane polyols using fraction primary, secondary, and hindered-secondary hydroxyl content. J. Appl. Polym. Sci. (2014)

85. D. Whisler, H. Kim, Experimental and simulated high strain loading response of polyurethane foam. J. Appl. Mech. **82**(4), 041005 (2015)

86. H.H. Al-Moameri, L.A. Jaf, G.J. Suppes, Simulation approach to learning polymer science. J. Chem. Educ. (2018)

87. R.G. Dingcong, D.B. Radjac, F.L.A. Alfeche, An iterative method for the simulation of rice straw-based polyol hydroxyl moieties. Sustainability **15**, 12082 (2023)

Chapter 2
Modeling Techniques

2.1 Theoretical Foundations for Thermo-kinetic Modeling

Due to the complexity of the chemical and thermal interactions in the synthesis of rigid polyurethane foam (RPUF), a thorough understanding of its theoretical foundations is crucial for accurate prediction. RPUFs have several applications which require materials of specific structural stability, thermal insulation, and durability. For this reason, thermo-kinetic modeling holds tremendous potential to advance not only academic research but also industrial production.

At the heart of RPUF synthesis are complex exothermic reactions involving polyols, isocyanates, and chemical blowing agents, which create polymeric structures and gas to constitute the foam. The chain of gelling and blowing reactions determines the foam's properties which include cellular structure, mechanical, and thermal properties. Simulations are especially valuable here, as they allow optimization of controlled parameters to optimize performance and anticipate challenges in scaling up production, without risking health and expending tremendous resources.

A thorough theoretical understanding is essential for thermo-kinetic simulations since this supports models that capture reaction kinetics, energy balance, and mass transfer. These robust models allow the resulting simulation to represent the dynamic nature of foam production. The following sections outline the general theoretical frameworks and equations needed to properly model and simulate the formation of RPUF. This work thus aims to further the predictive capability of simulations in RPUF by casting a firm foundation in transfer phenomena, thermodynamics, and reaction kinetics.

© The Author(s), under exclusive license to Springer Nature Singapore Pte Ltd. 2025   27
A. A. Lubguban et al., *Computational Thermo-kinetics of Rigid Polyurethane Foams*,
SpringerBriefs in Applied Sciences and Technology,
https://doi.org/10.1007/978-981-96-2077-7_2

2.1.1 Rigid Polyurethane Foam Kinetics

RPUF Reactions. Rigid polyurethane foams (RPUFs) are produced by complex reactions primarily involving isocyanates and polyols, typically catalyzed to control reaction rates and foam properties. The reactivity, kinetics, and type of reactions are important considerations for determining the final structural and functional properties of RPUFs. In this section, we discuss the main reactions contributing to polyurethane foam formation, focusing on the gelling and blowing reactions that play a central role in foam structure.

The gelling reaction or polymerization reaction, is the main pathway forming the polyurethane linkages. This reaction (Fig. 2.1) involves an isocyanate group reacting with a hydroxyl group in the polyol, resulting in a urethane linkage:

This polymerization step is crucial for the crosslinking that gives rigidity to the foam. Different types of hydroxyl groups, termed moieties, in polyols (primary, secondary, and hindered-secondary) impact the reactivity, with primary hydroxyl groups generally contributing to more compact networks, while secondary groups may introduce flexibility [1]. Furthermore, hybrid polyols, particularly amine-based polyols, also react with the isocyanate to produce polyurea besides polyurethane (Fig. 2.2):

Meanwhile, a critical component for foam formation is blowing agents. The blowing process can be performed chemically or physically, or through a combination of both methods. In a chemically blown RPUF, the blowing reaction is responsible for gas generation, which creates the foam's cellular structure (Fig. 2.3). This reaction occurs when isocyanate reacts with water to form a carbamic acid intermediate, which decomposes into an amine and carbon dioxide (CO_2). The generated CO_2

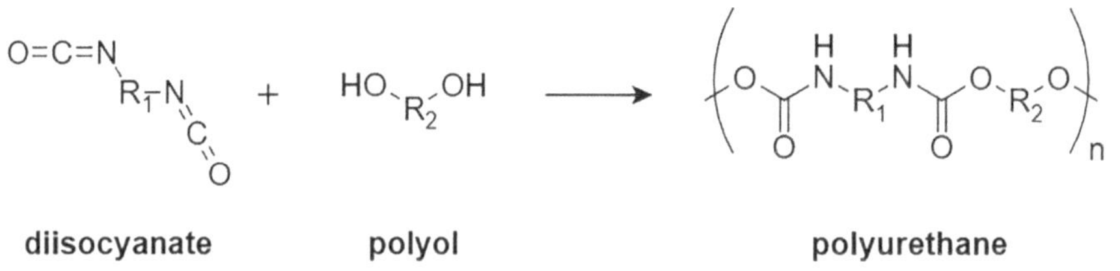

diisocyanate   **polyol**   **polyurethane**

Fig. 2.1 General gelling reaction for polyurethane

diisocyanate   **polyol**   **poly(urethane-urea)**

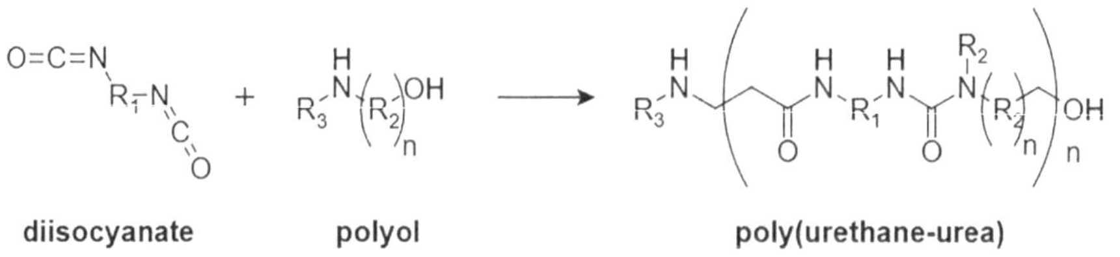

Fig. 2.2 General gelling reaction for poly(urethane-urea)

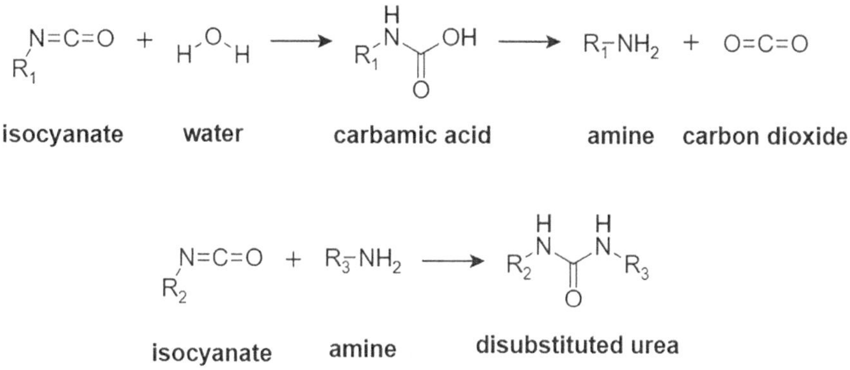

Fig. 2.3 Blowing reactions for RPUF

acts as a blowing agent, expanding the foam structure. This reaction is exothermic, contributing additional heat that accelerates other reactions within the system [2].

Aside from these reactions, side reactions may also occur during the polymerization process. Primary examples include biuret and allophanate linkages. These secondary reactions occur when isocyanate reacts with urea (producing biuret) or urethane (producing allophanate). Potential covalent crosslinking points are formed through these. However, they require higher temperatures and longer reaction times than the primary urethane and urea formation reactions, practically making their rate of formation negligible during RPUF reactions. Hence, they are often not considered in simulation studies [1, 2].

Temperature Dependence of the Rate Constant and the Rate Law. Taking into account these reactions, and several others, during the gelling and blowing processes, thermo-kinetic simulations for the synthesis of rigid polyurethane foams typically include Arrhenius-based kinetics. These kinetics describe the temperature dependence of the reaction rates of the polyurethane precursors—namely the polyol and isocyanate components. Equation 2.1 shows the Arrhenius equation, which can be used to determine the temperature dependence of the rate constant of the reactions, whether catalyzed or uncatalyzed.

$$k(T) = k_0 e^{\frac{-E_A}{RT}} \tag{2.1}$$

where

- k is the reaction rate constant;
- k_0 is the pre-exponential factor;
- E_A is the activation energy;
- R is the gas constant; and
- T is the absolute temperature.

As supported by extensive experimental data, it is common in the literature that polyurethane reactions are assumed to be elementary reactions that obey second-order kinetics [3]. Table 2.1 summarizes possible poly(urethane-urea) polymerization reactions and presents their corresponding reaction rates. As a distinction to simple molecules, the kinetics behind monomers are based upon the concentration of available moieties [4]. Additionally, the rate constants for each reaction may be determined, as shown in Eq. 2.2. Studies performed by Zhao et al. [4] show the catalyzed rate constant for each reaction i may be appended to the uncatalyzed counterpart as shown:

$$k_i = k_{i,\text{uncatalyzed}} + \Sigma\left(k_{i,\text{cat}\,j} \times \left[\text{cat}\,j\right]\right) \tag{2.2}$$

where

- $k_{i,\text{uncatalyzed}}$ is the reaction constant with no catalysis;
- $k_{i,\text{cat}\,j}$ is t the reaction constant with only catalyst j; and
- $\left[\text{cat}\,j\right]$ is the concentration of the catalyst j.

Viscosity-Dependent Pre-exponential Factor. In multiple simulation studies, the pre-exponential factor k_0 is assumed to be constant. Studies such as those performed by Ghoreishi et al. [5] successfully predicted the reaction temperature profiles by a method of assigning a fractional content of secondary hydroxyl group moieties which are hindered. However, Al-Moameri et al. [6] argued against considering hindered-secondary hydroxyl groups as another type of moiety. He reasoned that the observed slow reaction (previously attributed to hindered-secondary) was due to the rapid decrease in diffusion rates—as a result of increase in viscosity. For this reason, Al-Moameri et al. [6, 7] suggested using a viscosity-dependent frequency (pre-exponential) factor, accounting for mass transfer limitations. Using a group contribution method, the viscosity of polyols, isocyanates, and resulting polymer within the reaction matrix is estimated (Eq. 2.3).

$$\ln \frac{\mu_i}{\rho M} = A_i + \frac{B_i}{T} \tag{2.3}$$

where

- μ_i is the viscosity of polyol, isocyanate, or polymer;
- ρ is the density;
- M represents the molecular weight; and
- A_i and B_i are empirical constants determined for each component or group based on experimental data or group contribution values.

The viscosity data generated from the group contribution method feeds into a modified Arrhenius equation, where the frequency factor is adjusted based on viscosity changes. Assuming that the intermolecular diffusion is proportional to reaction temperature and inversely proportional to viscosity, and that the intramolecular diffusion is proportional to reaction temperature, the modified Arrhenius equation is

Table 2.1 Possible poly(urethane-urea) polymerization reactions adapted from Mendija et al. [1]

Reaction number	Reaction	Reaction rate expression
1	$A + B_p \rightarrow Ur$	$r_1 = k_1[A][B_P]$
2	$A + Bs \rightarrow Ur$	$r_2 = k_2[A][B_S]$
3	$A + B_H \rightarrow Ur$	$r_3 = k_3[A][B_H]$
4	$A + B_{SA} \rightarrow U$	$r_4 = k_4[A][B_{SA}]$
5	$A + B_P P \rightarrow Ur$	$r_5 = k_1[A][B_P P]$
6	$A + B_S P \rightarrow Ur$	$r_6 = k_2[A][B_S P]$
7	$A + B_H P \rightarrow Ur$	$r_7 = k_3[A][B_H P]$
8	$A + B_{SA} P \rightarrow U$	$r_8 = k_4[A][B_{SA} P]$
9	$AP + B_p \rightarrow Ur$	$r_9 = k_1[AP][B_P P]$
10	$AP + B_S \rightarrow Ur$	$r_{10} = k_2[AP][B_S P]$
11	$AP + B_H \rightarrow Ur$	$r_{11} = k_3[AP][B_H P]$
12	$AP + B_{SA} \rightarrow U$	$r_{12} = k_4[AP][B_{SA} P]$
13	$AP + B_P P \rightarrow Ur$	$r_{13} = k_1[AP][B_P P]$
14	$AP + B_S P \rightarrow Ur$	$r_{14} = k_2[AP][B_S P]$
15	$AP + B_H P \rightarrow Ur$	$r_{15} = k_3[AP][B_H P]$
16	$AP + B_{SA} P \rightarrow Ur$	$r_{16} = k_4[AP][B_{SA} P]$
17	$A + W \rightarrow Am$	$r_{17} = k_5[A][W]$
18	$AP + W \rightarrow Am$	$r_{18} = k_5[AP][W]$
19	$AP + Am \rightarrow U$	$r_{19} = k_6[AP][Am]$
20	$A + AmP \rightarrow U$	$r_{20} = k_6[A][AmP]$
21	$AP + AmP \rightarrow U$	$r_{21} = k_6[AP][AmP]$
22	$A + U \rightarrow P$	$r_{22} = k_7[A][U]$
23	$AP + U \rightarrow P$	$r_{23} = k_7[AP][U]$
24	$AP + UP \rightarrow P$	$r_{24} = k_7[AP][P]$

A mean isocyanate, B mean polyol, AP mean isocyanate moiety on (growing) polymer, BP mean polyol moieties on (growing) polymer, W mean water, Am mean primary amine, Ur mean urethane, U mean urea, UP mean urea in (growing) polymer, and AmP mean primary amine moiety on (growing) polymer. For the subscripts, P is primary alcohol moiety, S is secondary alcohol moiety, H is hindered-secondary alcohol moiety, and SA is secondary amine

as follows:

$$k = \left(A_1 \frac{T}{\mu} + A_2 T \right) e^{-\frac{E}{RT}} \tag{2.4}$$

where

- A_1 and A_2 are empirical parameters that depend on the specific reaction and material properties.

Chain-Growth and Step-Growth Polymerization. In thermoset polymerization, chain-growth and step-growth mechanisms are two unique pathways that determine how molecules assemble, directly influencing the final structure, properties, and speed of the polymerization process. Over time, our understanding of these mechanisms has deepened, especially with advances in catalysis and simulation methods that reveal the specific molecular steps involved in polymer growth. Al-Moameri and Jaf [8] recently performed extensive research on these pathways, investigating uncatalyzed and catalyzed reactions for RPUF formation.

The step-growth mechanism has been observed in non-catalytic systems or in the absence of specific active centers that accelerate reactions. In this mechanism, monomers react with one another in a gradual, pairwise manner, leading to an initially slow build-up of molecular weight. Each monomer has multiple functional groups, and it is assumed that their reactivity with others is equally probable. The polymerization progresses as monomers, dimers, trimers, and other low-molecular-weight species react stepwise, gradually increasing the molecular weight of the resulting polymer chains. Carothers [9, 10] originally described this mechanism and who observed that even at low extents of reaction, large molecular sizes could be achieved because of the availability of numerous functional groups on each monomer. This typically leads to the formation of a large number of relatively small oligomers in the initial stages, with high-molecular-weight polymers forming only at later stages. This incremental growth pattern means that the viscosity increase is slow, and the gel point is reached only once the reaction is close to complete.

Historically, the development of simulation approaches to capture step-growth kinetics has been significant in understanding how thermoset polymers crosslink into three-dimensional networks. As observed by Al-Moameri and Jaf [8], early theoretical work by Flory and Walling on gelation and crosslinking provided insights into how the gel point in step-growth reactions marks the transition from a liquid to a solid phase. This understanding has since been applied to simulate non-catalytic reactions in polyurethane systems, where accurate modeling of the step-growth pathway is essential for controlling resin viscosity and predicting the timing of gelation. Such simulations rely on parameters like reaction rate constants and molecular size distribution to monitor the gradual increase in molecular weight and guide formulation adjustments for thermoset products.

In contrast, the chain-growth mechanism is often associated with reactions that require catalysts to speed up the process. Here, polymerization happens by adding monomers one by one to specific active sites created by these catalysts. For example, in polyurethane production, catalysts interact with alcohol and isocyanate groups to create highly reactive points that allow the polymer chains to grow quickly, leading to high-molecular-weight polymers early in the reaction. As a result, chain-growth polymerization tends to produce fewer, larger polymer chains with high molecular weights, unlike the more gradual process seen in step-growth reactions.

Compared to step-growth polymerization, chain-growth reactions tend to cause a rapid increase in viscosity, as a few long-chain molecules quickly dominate the resin mixture. Simulation models—such as those using differential equations—can effectively capture the kinetics of these chain-growth reactions, especially in catalytic systems. These models allow manufacturers to predict when the material will start to gel and help design thermoset formulations with specific processing characteristics and mechanical properties.

2.1.2 Equations on Energy and Material

Thermodynamics in RPUF. Thermodynamics is crucial for understanding how rigid polyurethane foam (RPUF) is synthesized; it governs the energy and stability of the reactions involved. The foam forms through exothermic reactions between polyols and isocyanates. This released energy not only drives the reaction to completion but also affects the temperature, influencing both the rate of the reaction and the foam's final properties.

During the blowing reaction, thermodynamics also controls phase changes. For example, when physical blowing agents evaporate, or water (chemical blowing agent) reacts with isocyanate to release gases like carbon dioxide. The thermodynamic properties of the blowing agent, such as vapor pressure and heat of vaporization, influence the resins' behavior as temperature and pressure change during foam formation. The pressure generated by these gases within the foam cells is crucial, as it directly affects the foam's density, cell size, and overall structure.

Energy Balance in RPUF. Computational simulations of polyurethane foaming reactions and processes have been largely based on the work by Baser and Khakhar [11]. In this simulation study, authors performed an energy balance, assuming that the heat generated from the exothermic reactions is used for the blowing evaporation and as sensible heat of the reaction mixture/resin—an adiabatic process. Equation 2.5 shows a simplified form employed by [11].

$$\left(\Sigma n_i C_{p,i}\right)\frac{\mathrm{d}T}{\mathrm{d}t} = \frac{\Sigma(\Delta H_i r_i)}{\rho_p} - \lambda\left(-\frac{\mathrm{d}L}{\mathrm{d}t}\right) \tag{2.5}$$

where

- $\left(\Sigma n_i C_{p,i}\right)$ is the sum of the product of the mass per mass polymer and heat capacity of each of the following components: polymer, CO_2, blowing agent in liquid, and blowing agent in gas;
- $\Sigma(\Delta H_i r_i)$ is the sum of heat released during the gelling and blowing reaction;
- λ is the heat of vaporization of the blowing agent;
- ρ_p is the density of the polymer; and
- L is the mass of physical blowing agent.

However, while their research was the basis of more extensive studies on the topic, they failed to account for heat losses to the surrounding area and the effects of thermocouples on the resulting function; hence, modeling an isolated system. Tesser et al. [3] utilized a model that included heat transfer; authors also included an extended Flory–Huggins equation to deal with the vapor–liquid equilibrium of the blowing agent and the resin. Furthermore, succeeding authors such as Zhao et al. [2] and Ghoreishi et al. [5] performed an energy balance, nullifying the assumption of an adiabatic process. Energy losses to the surroundings via all modes of heat transfer may be lumped into an overall heat transfer equation, which is appended to the thermal energy balance. Assuming that the changes to the internal energy via the stretching of the polymer are negligible and that the heat of the reaction is constant, the final thermal energy balance is as follows:

$$\left(\Sigma n C_p\right)\frac{\mathrm{d}T}{\mathrm{d}t} = \Sigma(\Delta H_i r_i) - \lambda\left(-\frac{\mathrm{d}L}{\mathrm{d}t}\right) + UA\Delta T \tag{2.6}$$

where

- U is the overall heat transfer coefficient; and
- A is the heat transfer area.

Mass Transfer in RPUF. Mass transfer is a crucial part of making rigid polyurethane foam (RPUF), affecting how the foam expands, the structure of its cells, and the final qualities of the material. In RPUF formation, mass transfer refers to how chemical and physical blowing agents move from the liquid resin into the bubbles that form within the expanding foam. This movement drives the growth of bubbles, foam expansion, and ultimately creates the cellular structure that gives RPUF its insulating and mechanical properties.

RPUF relies on two types of blowing agents: chemical agents like water, which reacts with isocyanate to produce CO_2 gas, and physical agents, such as volatile hydrocarbons that evaporate due to the heat generated during the reaction. For the foam to expand evenly, the gases from both agents need to move efficiently from the liquid resin into the gas bubbles. The success of this mass transfer process is key to producing a well-structured foam with the desired characteristics.

The rate at which mass transfer happens depends significantly on the resin's viscosity, which changes as the reaction progresses. At first, when viscosity is low, blowing agents can move freely, allowing rapid and uniform bubble expansion. As the reaction continues, the resin thickens due to crosslinking, slowing down mass transfer. This slower diffusion limits bubble growth, setting the final foam structure and density. If mass transfer doesn't align with the reaction speed, blowing agents may get trapped in the resin, leading to smaller bubbles, less expansion, and a denser, less insulating foam.

Mass transfer also helps control the internal pressure within the foam. As gases enter the bubbles, the internal pressure rises, balancing the pressure of the surrounding resin and allowing stable bubble growth. If mass transfer is too slow, internal pressure may be too low, causing bubbles to collapse or form unevenly. On the other hand, a transfer rate that is too fast can create excessive pressure, potentially rupturing bubbles or making the foam overly expanded and weak. Achieving a balanced mass transfer rate is essential for stable, uniform cells that provide the foam with insulation and strength.

Overall Mass Transfer Rate, Rate of Vaporization, Bubble Growth. Mathematically, Fick's law describes the mass transfer of gases, according to which the mass flux is directly proportional to the infinitesimal changes to the gas concentration with respect to the boundary layer thickness. However, in RPUF formation, Fick's law poorly describes the diffusion of liquids, and it is a theory for gases. The diffusion of liquids is not well described in the literature, but it is mainly believed to be explained by hydrodynamics and activated states. Einstein and Smoluchowski's equation provides a way to calculate the liquid diffusivity by only using measurable parameters [12]. A simplified form can be derived assuming a spherical particle and the no-slip condition, as shown in Eq. 2.7:

$$D_{AB} = \frac{k_B T}{6\pi \eta a} \tag{2.7}$$

where

- D_{AB} is the liquid diffusivity;
- k_B is Boltzmann's constant;
- η is the dynamic viscosity of the resin; and
- a is the molecular radius of the blowing agent.

In the context of rigid polyurethane foam, the liquid diffusion coefficient is not constant. The reaction mixture (resin) experiences drastic temperature changes (up to 100 °C or more), and viscosity increases due to the gelling reactions, influencing the coefficient. Therefore, employing a temperature- and viscosity-dependent diffusion model is useful [12]. The general form of the overall mass transfer flux N, through a boundary layer, is given by Eq. (2.8).

$$N = K\Delta C \tag{2.8}$$

where

- K represents the overall mass transfer coefficient;
- A is the bubble surface area; and
- ΔC is the difference in the concentration between the gas and the liquid phases.

With further algebraic manipulation, a rate equation for the diffusion of the blowing agent from the liquid phase to the gas phase may be derived. Note, however, that this assumes all of the resistance is found in the liquid, the diffusion flux is approximately the bulk flux, and there is no loss of blowing agent. Equation 2.10 shows the rate of vaporization.

$$N = \frac{\frac{dn_g}{dt}}{A} = K\Delta C; \quad K = \frac{D_{AB}}{\Delta z} = \frac{k_B T}{6\pi \eta a \Delta z} \tag{2.9}$$

$$\frac{dn_g}{dt} = AK\Delta C = AK\left[\frac{-n_{PBA}}{V_{resin}}(\gamma x - x)\right] \tag{2.10}$$

where

- n_g is the moles of gas within the bubble;
- n_{PBA} is the initial moles of blowing agent in the liquid phase;
- V_{resin} is the volume of the resin;
- γ is the activity coefficient;
- x is the fraction of blowing agent in the gas phase; and
- x^* is the fraction of blowing agent in the liquid phase.

As the blowing agent evaporates, it causes the gas phase within the resin expands, forming bubbles that help the foam grow. Al-Moameri et al. [12] demonstrated that the rate of this vaporization (Eq. 2.10) may be derived using a modified Antoine equation to calculate the vapor equilibrium and the modified Raoult's Law to approximate the mole fraction of the liquid blowing agent at equilibrium. The activity coefficients were used as fitting parameters, assuming that they are a function of polymer concentration.

Meanwhile, the bubble growth may be determined using the modified Rayleigh-Plesset equation, adapted for the foam system to account for resin viscosity η, pressure differential across the bubble wall, and bubble radius r:

$$\frac{dr}{dt} = \frac{(P_g - P_L)}{4\eta\left(\frac{1}{r} - \frac{r^2}{S^3}\right)} \tag{2.11}$$

where

- P_g is the internal bubble pressure, which increases as blowing agents diffuse into the bubble;
- P_L is the pressure in the liquid reaction mixture;
- S is the outer radius of the resin layer that interfaces with the bubble; and
- η represents the dynamic viscosity of the resin, which increases over time as polymer crosslinking progresses.

Aside from bubble growth, the volume of the foam at different temperatures may also be determined. Zhao et al. [2] utilized the ideal gas law to arrive at a differential equation expressing the volume change with respect to reaction time:

$$\frac{dh}{dt} = \frac{dV}{dt} \times \frac{1}{A} = \left(\frac{22.4 \times T}{273.15}\right) \times \left(\frac{dn_{CO_2}}{dt} + \frac{dn_{PBA}}{dt}\right) \times \frac{1}{A} \tag{2.12}$$

where

- At standard conditions, the ideal gas equivalent volume is 22.4 L/mol; and
- $\frac{dn_{CO_2}}{dt} + \frac{dn_{PBA}}{dt}$ represent the rate of vaporization of the blowing agents.

2.2 Traditional Methods

2.2.1 Empirical Models

Empirical models have been foundational in rigid polyurethane foam (RPUF) research, providing a systematic approach to predict material properties based on experimental observations. By analyzing data obtained through controlled experiments, empirical models allow researchers to establish relationships between variables, helping to optimize RPUF properties for various applications, including insulation and structural support [13]. Unlike mechanistic models, which are grounded in theoretical principles, empirical models are data-driven and rely on statistical methods to correlate variables like density, compressive strength, and thermal conductivity. Empirical models take many forms, including linear, polynomial, and power-law equations, with the choice of model depending on the complexity of the relationship between the studied variables. One widely used empirical model in RPUF research links density with compressive strength, often represented by a power-law relationship:

$$\sigma_c = k \cdot \rho^n \tag{2.13}$$

where

- σ_c is the compressive strength;
- ρ is the density of the foam;
- k is a proportionality constant; and
- n is an exponent determined through regression analysis of experimental data.

This power-law model has proven particularly effective in RPUF research due to its ability to represent the nonlinear relationships often observed in foam materials, which have complex cellular structures that influence mechanical performance. Such models have been instrumental in developing RPUFs for applications requiring specific mechanical properties, particularly in load-bearing applications where compressive strength is critical. One of the major strengths of empirical models in RPUF research is their efficiency and adaptability to experimental data. Empirical models allow for quick predictions without requiring deep theoretical analysis, making them especially valuable in industrial contexts where timely decisions are crucial. In insulation applications, for example, empirical models enable rapid assessments of thermal conductivity based on foam density, streamlining material selection and production processes [14]. By relying directly on experimental data, empirical models can achieve high predictive accuracy within the conditions for which they were designed, as demonstrated by Kim et al. [14] in their study on RPUFs optimized for refrigeration. This study utilized empirical relationships to fine-tune foam density, thereby enhancing insulation properties without necessitating extensive iterative testing. Another advantage of empirical models is their applicability in quality control and experimental analysis. They enable researchers and engineers to establish correlations that help monitor consistency in material properties, facilitating checks on whether RPUF products meet the specified standards for insulation or structural performance. Empirical models have been especially useful in construction, where material properties like density and compressive strength are crucial for meeting safety and durability standards. By providing an empirical baseline, these models help manufacturers maintain quality throughout production [15].

However, despite these strengths, empirical models are inherently limited in their predictive power beyond the specific range of experimental conditions from which they are derived. This constraint is particularly significant when attempting to extrapolate results to novel formulations or untested environmental conditions, as empirical models lack the underlying mechanistic detail needed to adapt to unfamiliar scenarios. Zhao et al. [2] highlight that empirical models, while useful for capturing observed data trends, may not accurately predict behavior when foam composition or environmental conditions change significantly. For instance, a model developed to predict thermal conductivity based on density might not hold for a new RPUF formulation with altered cell structure or chemical composition, as these changes can impact thermal properties in ways the original model does not account for. The accuracy of an empirical model is heavily dependent on the quality and comprehensiveness of the data used to create it. Any inconsistencies or biases in the experimental data can lead to significant errors, as these inaccuracies are directly reflected in the model's predictions. This sensitivity to data quality means that empirical models are most reliable when derived from robust datasets that cover a broad range of conditions. Abdessalam et al. [13] discuss how limited or skewed data can reduce an empirical model's applicability, particularly if the dataset lacks representation at the extremes of the studied parameter range. As such, when using empirical models, researchers must carefully consider the quality and scope of their data to avoid overgeneralizing results.

Another drawback of empirical models is their lack of insight into the underlying mechanisms governing material behavior. Unlike theoretical or mechanistic models, which describe the physical principles driving a material's response, empirical models are simply a reflection of observed relationships. This limitation can impede material innovation, as empirical models alone cannot explain why specific changes in formulation lead to particular outcomes. Abdessalam et al. [13] emphasize that empirical models are most effective when combined with mechanistic approaches that provide a more comprehensive understanding of material performance.

The use of empirical models in RPUF research dates back to early studies that focused on understanding basic relationships between foam properties, such as density and compressive strength. These early models played a crucial role in establishing foundational knowledge that has shaped current practices in material optimization and applications. One of the first significant empirical models in RPUF research was the density-dependent compressive strength model, which helped demonstrate that higher-density foams generally exhibit greater compressive strength. Kim et al. [14] found that the denser cell structure of high-density foams provides greater resistance to deformation, making these foams more suitable for structural applications. The density-compressive strength model has had a lasting impact on RPUF material design, particularly for construction applications where foam panels must withstand substantial loads. By linking compressive strength to density, this model provided a framework for selecting foam densities that meet both insulation and load-bearing requirements. Raimbault et al. [15] note that these empirical relationships are still used in industry today, where they serve as a quick-reference tool for determining the suitability of RPUFs in different structural applications.

Another important empirical model in RPUF research addresses thermal conductivity, a critical property for insulation applications. Thermal conductivity (λ) is often related to foam density through a polynomial equation, such as:

$$\lambda = a + b \cdot \rho + c \cdot \rho^2 \tag{2.14}$$

where

- λ is the thermal conductivity;
- ρ is foam density; and
- a, b, c are constants derived from regression analysis.

This quadratic relationship has been used extensively to evaluate insulation performance in RPUFs of varying densities. Zhao et al. [2] applied this model in their study on RPUFs used in refrigeration units, where balancing thermal insulation with minimal material thickness is essential for energy efficiency. By using the thermal conductivity model, they optimized RPUF density to achieve the required insulation standards without overusing material. Today, empirical models often serve as a starting point for more complex modeling approaches, including hybrid models that integrate empirical data with theoretical insights for enhanced predictive accuracy. In the automotive industry, for instance, empirical models guide the selection of RPUF densities and compressive strengths for components designed to absorb impact. Weißenborn et al. [16] illustrate how empirical models are used alongside simulations, such as Finite Element Analysis (FEA), to refine material properties and meet stringent safety standards.

Despite their limitations, empirical models continue to be valuable in RPUF research. They allow for rapid predictions and enable researchers to analyze the effects of density and other variables on foam properties. Empirical models have paved the way for innovations in material formulation, making them indispensable in the development of RPUFs for diverse applications. As computational capabilities expand, the integration of empirical models with mechanistic and simulation-based methods will further enhance the versatility and accuracy of RPUF predictions.

2.2.2 Numerical Models

Numerical models have become indispensable in the study and optimization of rigid polyurethane foams (RPUFs). These models provide insights into how RPUFs behave under various mechanical and thermal conditions, enabling researchers and engineers to optimize foam properties for diverse applications. Among the most widely used numerical modeling techniques are Finite Element Analysis (FEA) and Finite Volume Method (FVM). FEA focuses primarily on the mechanical properties of RPUFs, allowing for the simulation of stress and strain responses under load, while FVM is primarily applied in thermal simulations to model heat transfer and thermal insulation performance. By leveraging these techniques, researchers have advanced RPUF applications in industries such as construction and automotive manufacturing, where material properties like compressive strength and thermal insulation are crucial.

Finite Element Analysis (FEA). Finite Element Analysis (FEA) is a powerful computational technique for simulating the mechanical behavior of materials. In RPUF research, FEA is used to predict how the foam will respond to various loading conditions, which is essential for applications requiring both structural integrity and energy absorption. The FEA process involves dividing the RPUF material into small, finite elements, each with defined material properties, and then applying mechanical forces to simulate stress, strain, and deformation.

The basic equilibrium equation in FEA for a linear elastic material under static conditions is:

$$Ku = F \tag{2.15}$$

where

- K is the global stiffness matrix of the material or structure, which depends on the material properties and the geometry of the finite elements;
- u is the displacement vector of nodes, representing the displacements at each node in the mesh; and
- F is the force vector applied to the structure, including external loads and boundary conditions.

This equation essentially balances the internal forces (captured by the stiffness matrix and displacements) with the external forces applied to the system.

The application of FEA in RPUF research allows engineers to simulate material responses under real-world conditions without physically testing each scenario, thus saving time and resources. FEA is particularly valuable in assessing how RPUFs perform under compression, tension, and impact conditions, providing a critical foundation for designing products that meet specific mechanical performance criteria. According to Weißenborn et al. [16], FEA has been instrumental in modeling the strain-rate-dependent behavior of RPUFs, helping engineers understand how foam density, cell structure, and other properties influence the material's overall performance.

Importance of Meshing, Boundary Conditions, and Material Properties in FEA. Developing an accurate FEA model involves several critical steps, with meshing, boundary conditions, and material property assignment being among the most important. Meshing divides the RPUF material into small elements, which allows for detailed analysis of how forces and deformations distribute across the foam. A finer mesh can improve model accuracy by providing more data points but increases computational requirements, while a coarser mesh reduces computational demand but can compromise accuracy. Selecting an appropriate mesh size is crucial for balancing these factors. Boundary conditions, which define the external forces and constraints on the material, are equally essential. These conditions simulate real-world scenarios by specifying how the foam is held in place, where loads are applied, and how it interacts with surrounding materials. Accurate boundary conditions ensure that the simulation reflects the actual operating environment, making the results more reliable. Weißenborn et al. [16] emphasize that improper boundary conditions can lead to significant errors, especially in applications where precise stress distribution is required, such as in automotive components designed for impact absorption. Material properties, including density, modulus of elasticity, and Poisson's ratio, are fundamental inputs in FEA modeling. Accurate representation of these properties ensures that the simulation reflects the true behavior of the RPUF under load. The material's density, for instance, directly affects its compressive strength and energy absorption capacity, which are critical in safety applications. Kim et al. [14] demonstrated how variations in material properties influence the performance of RPUFs used in refrigerator insulation, with FEA providing insights into how different foam densities affect mechanical and thermal stability.

FEA Applications. FEA has significantly contributed to optimizing RPUFs for construction and automotive applications. In the construction industry, RPUFs are widely used as insulation materials and structural support in wall panels, where they provide both thermal resistance and load-bearing capabilities. By using FEA, engineers can simulate how RPUF panels respond to compression and tensile forces, ensuring that they maintain structural integrity even under load. Raimbault et al. [15] highlight the role of FEA in designing RPUF panels with the optimal density and cell structure to withstand stresses in construction environments, improving both the durability and insulation performance of buildings. In the automotive sector, FEA is indispensable for designing RPUFs that enhance passenger safety by absorbing impact energy. RPUFs are used in automotive interiors and exteriors to reduce injury risk in collisions, with FEA simulations enabling engineers to optimize foam density and distribution for maximum energy absorption. According to Weißenborn et al. [16], FEA has been instrumental in predicting how RPUFs behave under high-strain rates, which is essential for applications like automotive crash safety. By simulating impact scenarios, engineers can refine the design of RPUF components to achieve the desired balance between weight and impact resistance, contributing to the development of safer, more efficient vehicles.

Finite Volume Method (FVM). While FEA focuses on mechanical behavior, the Finite Volume Method (FVM) is commonly used to model thermal properties in RPUFs. FVM is a numerical technique that divides the material into control volumes, allowing for the conservation of quantities such as heat and mass across each volume. This approach is particularly suited for thermal simulations, where accurate prediction of heat transfer and insulation performance is essential.

In RPUF applications, FVM enables engineers to model how heat moves through the foam under different environmental conditions. The method uses the heat conduction equation, which for RPUFs can be expressed as:

$$\frac{\partial}{\partial t}\left(\rho c_p T\right) = \nabla \cdot (k \nabla T) + Q \qquad\qquad (2.16)$$

where

- T is the temperature;
- ρ is the density;
- c_p is the specific heat capacity;
- k is the thermal conductivity; and
- Q represents internal heat sources or sinks.

This equation allows for the simulation of both steady-state and transient heat transfer, making FVM ideal for assessing RPUF thermal insulation performance in applications such as building insulation and refrigeration.

Accurate thermal modeling in FVM requires setting appropriate boundary conditions, including fixed temperature (Dirichlet), heat flux (Neumann), and convective (Robin) boundary conditions, depending on the environment in which the RPUF is used. Kim et al. [14] applied FVM to model thermal behavior in RPUFs for refrigerator insulation, demonstrating how boundary conditions influence heat flow and temperature gradients. Such studies help engineers predict how insulation properties vary with temperature, enabling the design of foams that maintain thermal resistance over time.

FVM Applications. Thermal management is a critical consideration in both construction and refrigeration, where RPUFs serve as primary insulation materials. In construction, RPUFs are used to insulate walls, roofs, and other structures, helping to reduce energy consumption by limiting heat transfer. FVM is applied to model the heat flow through RPUFs over time, taking into account factors like density and material composition. In regions with extreme temperatures, these simulations help engineers design RPUF panels that maintain thermal resistance under environmental stress, as shown in Zhao et al. [2], who used FVM to predict long-term insulation performance in building materials. In refrigeration, where energy efficiency is essential, FVM simulations help optimize foam properties to achieve high insulation with minimal thickness. Kim et al. [14] demonstrated that FVM models could predict temperature gradients within RPUF panels, enabling engineers to adjust material parameters for enhanced insulation performance. This approach has contributed to the development of more energy-efficient refrigerators by identifying the optimal density and thermal conductivity for RPUF insulation.

Case Studies. The application of FEA and FVM in industrial settings demonstrates the versatility of these numerical techniques in optimizing RPUF properties. Each method has contributed to advancements in construction, automotive, and refrigeration industries, providing insights that guide product design and material selection. In the construction industry, FEA and FVM have been applied to develop RPUF panels that offer both structural support and thermal insulation. Using FEA, engineers can model the compressive strength of RPUFs, ensuring they can withstand the weight of other building materials and environmental loads. Simultaneously,

FVM allows for the assessment of thermal performance, helping to design insulation panels that reduce energy transfer and improve building efficiency. Raimbault et al. [15] discuss how combining FEA and FVM in RPUF panel design has resulted in materials that are both strong and thermally resistant, meeting the demands of modern construction.

In automotive applications, FEA is particularly valuable for enhancing crash safety through the design of impact-absorbing components. Weißenborn et al. [16] illustrate how FEA simulations can predict how RPUFs deform under impact, allowing engineers to adjust density and cell structure to maximize energy absorption. By optimizing RPUF properties, FEA has enabled the development of lighter, safer vehicles that meet rigorous safety standards. Similarly, in automotive thermal management, FVM models are used to predict how RPUF insulation behaves under engine heat, ensuring that interior compartments remain insulated from high temperatures [14].

Despite the successes of FEA and FVM in industrial applications, challenges remain. Numerical models often require extensive experimental validation to ensure accuracy, especially when simulating complex foam behaviors. Models must be calibrated with experimental data to capture specific material properties accurately, and errors in mesh design or boundary conditions can lead to significant inaccuracies. Zhao et al. [2] and Abdessalam et al. [13] both emphasize the importance of experimental validation in RPUF simulations, as discrepancies between model predictions and real-world performance can lead to design failures or material inefficiencies.

2.3 Modern Techniques

2.3.1 Database-Driven Predictions

The advent of large datasets has significantly transformed the landscape of material property modeling, particularly in the context of rigid polyurethane foams (RPUFs). These materials are widely recognized for their excellent thermal insulation and mechanical properties, making them invaluable in various applications, including construction and automotive industries. Access to extensive material property databases is a foundation for machine learning (ML) applications in this field. These databases contain vast data on various material properties, including thermal conductivity, density, and mechanical strength, which are essential for training predictive models specific to RPUFs. For instance, various databases in Table 2.2 include over 139,367 crystalline compounds, providing a rich resource for developing and validating ML algorithms to predict the behaviors of materials like RPUFs.

The integration of large datasets into modeling frameworks allows researchers to leverage sophisticated machine learning techniques, such as gradient boosting regression (GBR) and artificial neural networks (ANNs), which have demonstrated success in predicting material properties [18]. The availability of extensive data

Table 2.2 Commonly used databases in material sciences based on crystalline structures and properties [17]

Database	Structure	Properties
Cambridge structural database (CSD)	1,031,632	
Inorganic crystal structure database (ICSD)	218,839	
Crystallography open database (COD)	457,771	
International center for diffraction data (ICDD)	1,004,568	
AFLOW	3,249,264	Formation energy, band structures, and Bader changes
		Elastic and thermal properties
		Binary, ternary, and quaternary systems
Materials Project	654,758	Band structures
		Elastic and piezoelectric tensors
		Porous volume and surface
Open quantum materials database (OQMD)	637,644	Formation energy and band structures

enables these models to learn complex relationships between the structure and properties of RPUFs, thereby improving the accuracy of predictions. For example, recent studies have shown that using large-scale computational databases can significantly enhance the performance of models designed to predict the thermal and mechanical properties of rigid polyurethane foams based on their chemical composition and processing conditions [19]. This is particularly important in materials science, where the relationships between structure and properties can be intricate and nonlinear.

Moreover, the quality of datasets cannot be overstated. High-quality datasets that are comprehensive and representative of the material space allow for better generalization of the models trained on them. Research indicates that the accuracy of predictions is heavily influenced by the quality and heterogeneity of the datasets used [20]. In scenarios where large datasets contain noise or inconsistencies, advanced techniques such as transfer learning can mitigate these issues, allowing models to perform well even with imperfect data. This adaptability is crucial in materials science, where experimental data on RPUFs can be limited or challenging to obtain.

Additionally, using large datasets facilitates the exploration of new formulations and the discovery of novel formulations in RPUFs. For example, machine learning approaches have successfully accelerated the discovery of RPUFs with enhanced thermal insulation properties by optimizing the chemical composition and processing parameters [21]. By mapping the relationships between synthetic variables and material properties, researchers can utilize machine learning to predict outcomes for RPUFs that have yet to be synthesized, thus streamlining the material design process.

Machine learning focuses on developing algorithms that autonomously learn from data to create statistical models for analysis and prediction, enabling accurate outcomes without explicit programming for specific tasks. Recent years have witnessed rapid advancements in ML applications, extending beyond computer scientists to researchers across various fields, including chemical and materials sciences, where ML accelerates computational tasks and addresses challenges that traditional modeling methods struggle to solve. Deep learning, a subset of ML based on ANNs, further enhances these advancements by enabling more complex data analysis and pattern recognition. This chapter aims to illustrate the general workflow (Fig. 2.4) of machine learning within the computational material's context, providing examples and considerations for implementing ML-based modeling. We will highlight several successful machine learning techniques applied in computational chemistry and materials science, including recent developments in deep learning while acknowledging that our discussion is not exhaustive. For a comprehensive review of machine learning in molecular and materials science, we recommend the thorough introductory work by Butler et al. [22].

Data Gathering. ML algorithms are trained on existing datasets to learn and enhance their predictive capabilities, making the size and quality of these datasets crucial for developing accurate models. The initial step in the ML workflow involves identifying, gathering, or creating training datasets, which are heavily influenced by the specific goals of the model being developed. Researchers can access free

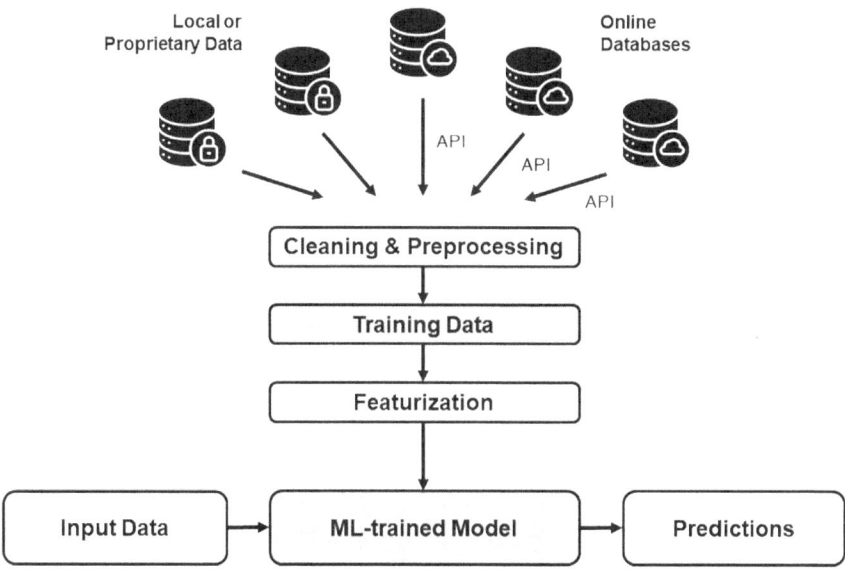

Fig. 2.4 General machine learning workflow for material property prediction. Adapted from "Machine learning approaches for the prediction of materials properties," by Y. Chibani and F.-X. Coudert, published in APL Materials, 2020, https://doi.org/10.1063/5.0018384, under the CC BY 4.0 license

datasets from platforms like Kaggle, the UC Irvine Machine Learning Repository, or various governmental open-data initiatives for general purposes. In materials science, numerous validated datasets are publicly available, although some may have stricter licensing terms or require subscriptions. These datasets primarily fall into two categories: those focusing on structural information and those emphasizing physical or chemical properties of materials. The first category of databases, which focuses on the structures of materials, is among the most established and widely recognized databases in material science in the field, as shown in Table 2.2. A prominent example is the Cambridge Structural Database (CSD), which houses over one million experimental crystal structures and serves as a standard repository for new crystal structures, including organic and metal–organic compounds [23]. Other notable databases include the Inorganic Crystal Structure Database (ICSD) for inorganic crystals and the Crystallography Open Database (COD), which promotes open access to crystallographic data [24]. Additionally, databases such as the GDB for small organic molecules and ZINC for commercially available compounds facilitate virtual screening [25]. These resources are invaluable for researchers, providing a wealth of structural information that can be utilized in various applications, from computational modeling to materials discovery. The availability of these databases, often with open access, significantly enhances the ability of scientists to conduct research and innovate in materials science [26].

It is essential to recognize a significant bias inherent in the databases available in the field of materials science: they predominantly focus on crystallographic structures. This limitation is tied to the nature of structural determination and representation, which can skew the overall understanding of materials science. Such biases, whether implicit or explicit, arise from the specific scope and representation choices made by these databases, which may not encompass the full breadth of materials science research [27]. In recent years, however, there has been a notable increase in the development of structure–property databases, which provide extensive data often centered on specific classes of materials or properties. These databases, many of which are open-access and collaborative, present valuable opportunities for training and validating new ML models [28]. Notable examples include the Materials Project, which focuses on inorganic materials, the Automatic Flow for Materials Discovery (AFLOWLIB), and the Open Quantum Materials Database (OQMD) [29]. Additionally, specialized databases such as the Harvard Clean Energy Project for organic solar materials and the NREL Materials Database for renewable energy materials further illustrate the diversity of available resources [30].

Most of these databases offer user-friendly web interfaces for exploration and visualization, as well as Application Programming Interfaces (APIs) that facilitate automated access to data. APIs are designed to provide machine-readable queries and results, making them suitable for integration into various projects. For instance, the Python Materials Genomics (pymatgen) package integrates seamlessly with the Materials Project's RESTful API, enhancing accessibility for researchers [31]. Moreover, high-throughput calculations can generate datasets "on the fly," allowing researchers to create ML training sets that can be published alongside their findings or submitted to online repositories [32].

The emergence of these diverse databases, coupled with the ability to generate new data through computational methods, is transforming the landscape of materials science. By providing a more heterogeneous dataset, these resources enable researchers to explore the relationships between structure and properties more effectively, ultimately accelerating materials discovery and innovation. However, it remains crucial for researchers to be aware of the biases present in the datasets they utilize, as these can significantly influence the outcomes of their analyses and the universality of their models. As the field continues to evolve, integrating data-driven approaches with traditional experimental methods will likely yield new insights and advancements in materials science, paving the way for future innovations.

Data Processing. Using existing datasets of material structures and properties or generating new datasets is critical for effective ML applications in materials science. However, these datasets cannot be directly integrated into ML workflows in their original formats. When dealing with large datasets, four key characteristics, known as the "four V's," must be considered: volume, variety, veracity, and velocity. Volume refers to the sheer amount of data available; variety pertains to the heterogeneity of the data in terms of format and meaning; veracity involves understanding the uncertainties associated with each data point; and velocity indicates the speed at which data is generated and processed, although this is less of a concern in many material science workflows, which do not typically require real-time processing.

Before datasets can be effectively utilized in ML models, they must undergo a process of homogenization and cleaning. This involves identifying erroneous, missing, or inconsistent data points—often outliers—using criteria grounded in physical or chemical principles. This data curation is essential for building accurate predictive models, as the presence of outliers can significantly skew results. The necessity for data cleaning can vary depending on the specific ML algorithms employed; for instance, some algorithms, such as those in the Random Forest family [33], cannot handle null values, while others may be more robust in such inconsistencies.

A pertinent example of the importance of dataset curation can be drawn from a recent large-scale analysis of the elastic properties of inorganic crystalline materials sourced from the Materials Project database [34]. This study found that out of 13,621 crystals analyzed, only 11,764 were mechanically stable, while 1857 (approximately 14%) exhibited elastic tensors indicative of mechanical instability, rendering them unsuitable for further analysis. Additionally, some materials presented elastic moduli that, while mathematically valid, were unphysically large and required exclusion from the dataset. The development of structure–property databases has proliferated in recent years, providing a wealth of data that is often open access and collaborative. These databases, such as the Materials Project, AFLOWLIB, and the Open Quantum Materials Database, offer significant opportunities for training and validating new ML models. Most of these databases are accessible via user-friendly web interfaces and APIs, facilitating automated data retrieval and integration into various projects. For example, the pymatgen package allows seamless interaction with the Materials Project's RESTful API, enhancing data accessibility for researchers.

Data Presentation. Once the data have been cleaned and homogenized, the subsequent step in the machine learning (ML) workflow involves encoding this data into a set of specific variables that the ML algorithm can manipulate. Raw data often require transformation into a suitable format for the learning process, typically represented as scalar or vector variables for each entry in the dataset. This transformation may include techniques such as rescaling, normalization, or binarization to ensure the data are in a state the algorithm can efficiently process. Standard preprocessing methods, such as MinMaxScaler and StandardScaler, available in libraries like scikit-learn, facilitate this step [35]. However, it is crucial to carefully evaluate the impact of these preprocessing techniques; in some cases, algorithms may perform better without extensive preprocessing, as excessive alteration can obscure critical features essential for optimal performance.

When the input data consists of chemical structures, the choice of representation becomes particularly challenging. Chemical compounds and materials are inherently complex three-dimensional entities, and representing them directly as vectors of coordinates may not be the most effective approach for ML workflows. This challenge is addressed through a process known as featurization or feature engineering, which is an active area of research focused on developing optimal representations of chemical structures. The primary objectives of feature engineering are twofold: first, to prepare input data in a manner conducive to the specific characteristics of the chosen ML algorithm, and second, to enhance the performance of ML models by leveraging chemical intuition to identify and encode essential features of the materials.

Chemical information can be transformed into a series of descriptors that encapsulate essential characteristics of the dataset, allowing the ML algorithm to train effectively. Various mathematical representations serve as descriptors for chemical and material structures, including the Coulomb matrix, Simplified Molecular Input Line Entry System (SMILES), bag of bonds, molecular graphs, and BAML (bonds, angles, machine learning) [36]. For crystalline structures, common representations include translation vectors, fractional coordinates of atoms, radial distribution functions, Voronoi tessellations of atomic positions, and property-labeled material fragments [37].

The choice of descriptor significantly influences the success of ML applications in materials science. For instance, the development of graph neural networks (GNNs) has enabled the learning of material properties directly from graph-like representations of crystal structures, providing a versatile and accurate framework for ML [38]. Additionally, recent advances in deep learning architectures, such as SchNet, have demonstrated the ability to recover fundamental chemical knowledge from datasets of bulk crystals, showcasing the potential of these methods to yield fast and accurate predictions [39].

Model Development and Evaluation. The next step in the ML workflow is training the ML algorithm, utilizing the curated and pre-processed dataset as input. ML models can be broadly categorized into supervised, unsupervised, and semi-supervised. In supervised learning, the dataset serves as a training set that includes both input variables and their corresponding output variables. A common example in chemistry is the relationship between chemical structures and their properties. The primary objective of the ML algorithm in this context is to learn the mapping function from the input (the structures) to the output (the properties). Once trained, the algorithm should be capable of making accurate predictions for new data. Supervised learning problems can be divided into two main categories: regression and classification techniques. Both aim to construct a model that predicts values based on available variables, with the distinction of the output variable. The predicted variable is continuous in regression tasks, such as melting point, bandgap, or elastic modulus. Common regression algorithms include linear regression, lasso regression, ridge regression, elastic net regression, and Gaussian process regression. Linear regression (LR) is the simplest of these algorithms, aiming to create the best linear model based on the provided descriptors. Lasso regression can be employed for more complex models, which modifies the loss function to minimize model complexity.

In contrast, classification problems involve predicting categorical variables, where the algorithm assigns labels to input data. The simplest classification scenario is binary classification, which determines whether a material is conductive or insulating. Numerous classification algorithms are available, including logistic regression, linear discriminant analysis, k-nearest neighbors, naïve Bayes, classification and regression trees, support vector machines, and kernel ridge regression (KRR). For instance, Ghiringhelli et al. utilized KRR with descriptors derived from energy levels and radii of valence orbitals to predict crystalline arrangements between zinc blende and wurtzite structures [40].

Unsupervised learning operates differently by drawing inferences from input data without corresponding output variables. This approach is particularly useful for uncovering previously undetected patterns in data with minimal human supervision. Unsupervised learning can analyze unlabeled data to discover inherent groupings, with the ML algorithm identifying trends that rationalize the dataset and present it in novel ways. A key family of unsupervised learning algorithms is clustering or cluster analysis, which divides data into groups (or clusters) based on similar features without prior assumptions about the nature of these groups. Classical clustering methods include Gaussian mixtures, k-means, hierarchical, and spectral clustering. Other unsupervised learning methods include association rule learning and principal component analysis, which establish relationships among multiple features in large datasets, a task that is often challenging to perform manually. Supervised learning is more prevalent in chemistry and materials science, although there are notable examples of unsupervised learning applications. For instance, Saad et al. employed supervised and unsupervised techniques to predict the structure and properties of binary compounds, such as melting points [41]. Supervised ML models have been trained to reproduce the lowest unoccupied molecular orbital (LUMO) and highest occupied molecular orbital (HOMO) for organic solar cells [42] and to predict thermodynamic parameters like adsorption energy and activation energy in catalytic processes [43]. Other Studies have developed predictive Quantitative Materials Structure–Property Relationship (QMSPR) models to forecast critical temperatures of known superconductors [44]. At the same time, Woo and colleagues established a QMSPR model for high-throughput screening of metal–organic frameworks (MOFs) for CO_2 capture through adsorption [45].

One limitation of supervised ML algorithms is acquiring labeled data, which can be costly and time-consuming, especially with large datasets. In contrast, unlabeled data are relatively inexpensive and easier to collect. Consequently, applications of unsupervised learning in materials science have been limited. An alternative approach is semi-supervised learning, which combines supervised and unsupervised learning elements. In this framework, a large amount of input data is available, but only a limited amount of corresponding output data exists. Semi-supervised learning aims to leverage the labeled portion of the dataset to train an accurate model. Initially, unsupervised learning algorithms cluster similar data and apply supervised techniques to predict outcomes for the remaining unlabeled data. Semi-supervised learning has applications in text and speech analysis, internet content classification, and protein sequence classification. For example, Court et al. utilized semi-supervised learning to create a materials database of Curie and Néel temperatures for 39,822 compounds by text mining a corpus of 68,078 chemistry and physics articles, employing natural language processing and a semi-supervised relationship extraction algorithm to extract property values [46]. Similarly, Huo et al. [47] demonstrated the effectiveness of semi-supervised learning in classifying inorganic materials synthesis procedures from natural language descriptions. Recently, Kunselman et al. [48] applied semi-supervised learning methods to analyze and classify microstructure images, training their model on a dataset with only a fraction of labeled microstructures.

2.3.2 *Iterative Predictions*

The iterative approach to model refinement has emerged as a modern technique for enhancing the predictive capabilities of ML models in materials science, particularly for applications involving RPUFs. This process involves continuously improving predictions through feedback from experimental results, creating a dynamic loop of learning and adaptation. The iterative cycle in Fig. 2.5 typically begins with an initial model trained on a curated dataset and then evaluating its predictions against experimental data. Discrepancies between predicted and observed values provide critical insights that inform subsequent iterations of model development.

The iterative process can be particularly beneficial due to the complex interplay of factors influencing their properties, such as chemical composition, processing conditions, and environmental factors. For instance, researchers may initially develop a predictive model for the thermal conductivity of RPUFs based on a limited set of experimental data. As new experimental results become available, these can be integrated into the dataset, allowing for model retraining and refinement. This feedback loop enhances the model's accuracy and enables the identification of previously unconsidered variables that may influence material properties.

The iterative process typically involves several key steps:

1. **Model Development**: The iterative RPUF thermo-kinetic model starts with a recipe that stores the initial input parameters of all physicochemical factors, including the heat capacities of all components, OH number, moieties, molecular weight, the functionality of polyols, density of polyols, and the Isocyanate index used, which is

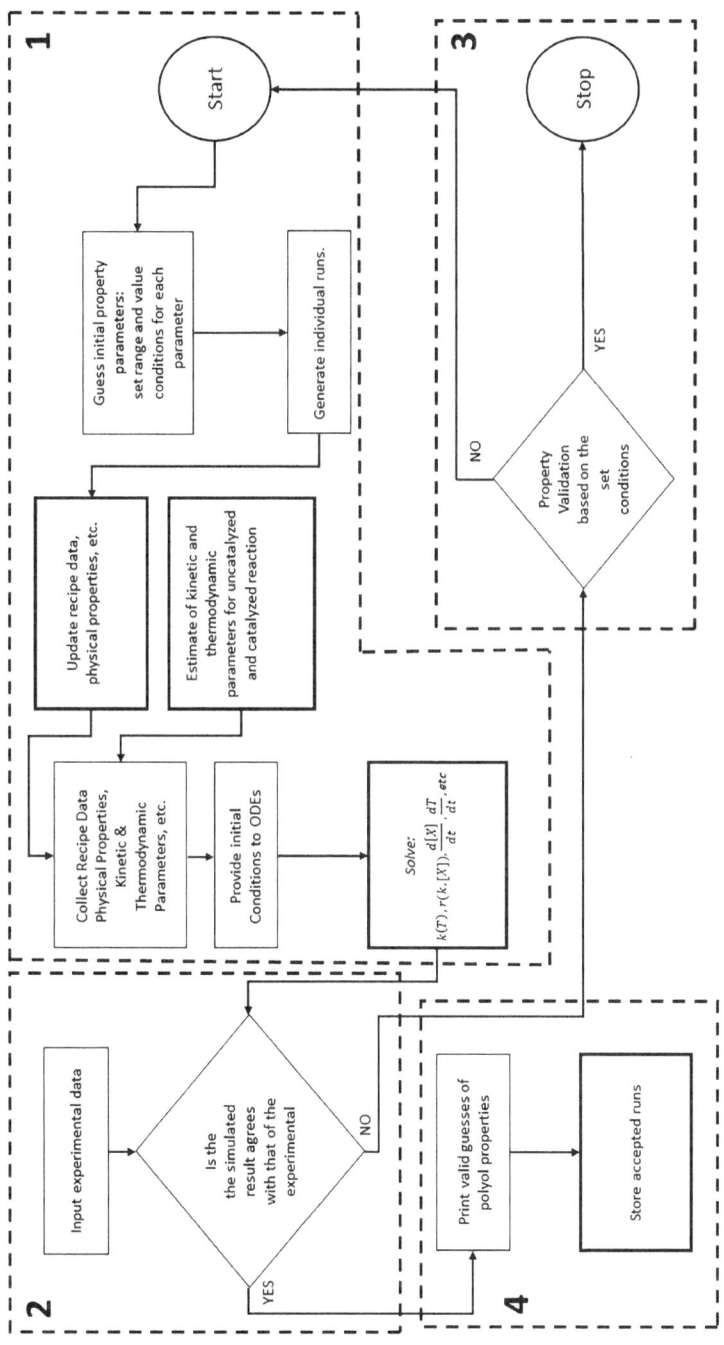

Fig. 2.5 General algorithm for the rigid polyurethane foam modeling using iterative method. Reproduced from Dingcong et al. [49]. Under the CC BY 4.0 license

defined as the ratio of the equivalent amount of isocyanate used relative to the theoretical equivalent of polyol times 100. Additionally, conditions such as ranges and extremities must be defined in the recipe. For example, the total value of all hydroxyl moieties should equal to 1. Moreover, bootstrapping can be done for each parameter combination to generate a simulated thermo-kinetic model. The modeling can be done by expressing the generated heat during both blowing and gelling reactions (Table 2.3) as the solution of the differential thermodynamic equation in the form of internal temperature vs time as expressed in Eq. 2.17 [2].

$$\frac{\mathrm{d}T}{\mathrm{d}t} = \frac{\sum_i \Delta H_{(gel)i} \times r_{(gel)i} + \Delta H_{(blow)i} \times r_{(blow)i} + UA\Delta T}{\sum (n \times C_p)} \qquad (2.17)$$

where

Table 2.3 Possible reactions involved in poly(urethane-urea) polymerization include various components: A denotes isocyanate, B represents polyol, AP refers to the isocyanate group on a growing polymer, and BP indicates polyol groups on a growing polymer. Other components include W for water, Am for primary amine, Ur for urethane, U for urea, UP for urea within the growing polymer, and AmP for the primary amine group on the growing polymer. Subscripts are used to specify alcohol types: P denotes primary alcohol, S is secondary alcohol, H is hindered-secondary alcohol, and SA refers to secondary amine

Reaction No	Reaction	Reaction rate expression
1	$A + B_p \rightarrow Ur$	$r_1 = k_1[A][B_P]$
2	$A + B_S \rightarrow Ur$	$r_2 = k_2[A][B_S]$
3	$A + B_H \rightarrow Ur$	$r_3 = k_3[A][B_H]$
4	$A + B_{SA} \rightarrow U$	$r_4 = k_4[A][B_{SA}]$
5	$A + B_PP \rightarrow Ur$	$r_5 = k_1[A][B_PP]$
6	$A + B_SP \rightarrow Ur$	$r_6 = k_2[A][B_SP]$
7	$A + B_HP \rightarrow Ur$	$r_7 = k_3[A][B_HP]$
8	$A + B_{SA}P \rightarrow U$	$r_8 = k_4[A][B_{SA}P]$
9	$AP + B_P \rightarrow Ur$	$r_9 = k_1[AP][B_PP]$
10	$AP + B_S \rightarrow Ur$	$r_{10} = k_2[AP][B_SP]$
11	$AP + B_H \rightarrow Ur$	$r_{11} = k_3[AP][B_HP]$
12	$AP + B_{SA} \rightarrow U$	$r_{12} = k_4[AP][B_{SA}P]$
13	$AP + B_PP \rightarrow Ur$	$r_{13} = k_1[AP][B_PP]$
14	$AP + B_SP \rightarrow Ur$	$r_{14} = k_2[AP][B_SP]$
15	$AP + B_HP \rightarrow Ur$	$r_{15} = k_3[AP][B_HP]$
16	$AP + B_{SA}P \rightarrow Ur$	$r_{16} = k_4[AP][B_{SA}P]$
17	$A + W \rightarrow Am$	$r_{17} = k_5[A][W]$
18	$AP + W \rightarrow Am$	$r_{18} = k_5[AP][W]$
19	$AP + Am \rightarrow U$	$r_{19} = k_6[AP][Am]$
20	$A + AmP \rightarrow U$	$r_{20} = k_6[A][AmP]$
21	$AP + AmP \rightarrow U$	$r_{21} = k_6[AP][AmP]$
22	$A + U \rightarrow P$	$r_{22} = k_7[A][U]$
23	$AP + U \rightarrow P$	$r_{23} = k_7[AP][U]$
24	$AP + UP \rightarrow P$	$r_{24} = k_7[AP][P]$

- U is the overall heat transfer coefficient with respect to the surroundings;
- A represents the foam's surface area;
- $\Sigma_i \Delta H_{(gel)i}$ describes the instantaneous heat released by the reactions occurring during the gelling and blowing processes;
- $UA \Delta T$ indicates the instantaneous heat transfer from the system to the surroundings; and
- $\sum (n \times C_p)$ represents the instantaneous heat capacity of the mixture.

Moreover, the gelling and blowing reactions are kinetically expressed in Eq. 2.18.

$$r_{gel} = \sum_i k_{gel,i} \times C_{catgel} \times C_{iso} \times C_{OH,i} = \sum_i r_{gel,i} \qquad (2.18)$$

$$r_{blow} = \sum_i k_{blow,i} \times C_{catblow} \times C_{iso} \times C_{BA \text{ functionality}, i} = \sum_i r_{blow, i} \qquad (2.19)$$

where

- r_{gel} and r_{blow} represent the total reaction rate of gelling and blowing mixtures;
- $k_{gel, i}$ and $k_{blow, i}$ are the reaction rate constants;
- C_{catgel} and $C_{catblow}$ denote the concentration of gelling and blowing catalysts;
- C_{iso} represents the concentration of the isocyanate used; and
- $C_{OH,i}$ and $C_{BA \text{ functionality}, i}$ represent the concentrations of the hydroxyl groups in the polyol and the functional groups present in the blowing agent used (e.g., $C_{OH,i}$ if water is used as the blowing agent).

Furthermore, Zhao et al. [2] reported an accurate expression of the catalyzed reaction rate constants as defined in Eq. 2.20.

$$k_i = k_{uncat,i} + \Sigma \left([cat] \times k_{cat,i} \right) \qquad (2.20)$$

where

- k_i denotes the overall reaction rate constant for gelling and blowing reactions in Table 2.3; and
- $k_{uncat,i}$ and $k_{cat,i}$ represent the reaction rate constants of uncatalyzed and catalyzed reactions, respectively.

The values of these rate constants may also vary at different temperatures, $k(T)$, as defined by the Arrhenius equation (Eq. 2.21).

$$k(T) = k_0 \times e^{\frac{-E_a}{RT}} \qquad (2.21)$$

where

- k_0 is the pre-exponential factor;
- E_a is the Activation energy; and
- R denotes the universal gas constant.

Equation 2.17 can be solved using a database or the bootstrapping output of the relevant physical, thermodynamic, and kinetic parameters, along with initial condition data. However, this process may be time-consuming and complex, as it involves solving multiple differential equations, considering all possible reactions (Table 2.3) simultaneously occurring during the process. In addition, the reaction rate constants vary for different hydroxyl moieties, as indicated by their uncatalyzed relative reaction rates (Table 2.4).

Consequently, each type of moiety exhibits different thermo-kinetic parameters when reacting, summarized in Table 2.5. Hence, a stepwise process evaluation should be studied for every gelling and blowing reaction type. Since these reactions are highly influenced by the type of polyols used, several studies have been conducted to tailor thermo-kinetic models precisely to the polyol type and elucidate the impact of various parameters.

2. **Prediction and Curve-fitting**: The predictive capability of a developed model is critically assessed through its application to a validation dataset or newly acquired experimental data. This process involves the evaluation of the model's performance through curve-fitting methodologies, which quantitatively measure the accuracy of the model's predictions against observed data. Commonly employed metrics for this evaluation include the Mean Absolute Error (MAE) and the Root Mean Square Error (RMSE). These statistical measures provide insight into the model's predictive fidelity, with lower values indicating a closer fit to the empirical data. It is worth noting that the validation of any predictive model is contingent upon the availability of a well-established database that encompasses thermo-kinetic parameters. This database is a foundational reference for the model, allowing for integrating

Table 2.4 Relative rates of isocyanate reaction against different hydrogen-active compounds

Hydrogen-active compound	Formula	The relative reaction rate (Non-catalyzed, 25 °C)
Primary aliphatic amine	$R-NH_2$	2.50×10^3
Secondary aliphatic amine	R_2-NH	500–1250
Primary hydroxyl	$R-CH_2-OH$	2.50
Water	HOH	2.50
Secondary hydroxyl	$R_2-CH-OH$	0.750
Tertiary hydroxyl	R_3-C-OH	0.0125

Table 2.5 Thermo-kinetic parameters of the catalytic reactions of isocyanate with primary, secondary, and hindered-secondary hydroxyls [21]

	Pre-Exponential factor (k_0)	Activation energy (Ea), J/mol	Heat of reaction (ΔH), J/mol
Primary hydroxyl	500	37,000	68,000
Secondary hydroxyl	55	40,000	68,000
Hindered-secondary Hydroxyl	42	40,000	68,000

experimental inputs such as the specific formulation (recipe) and the temperature profiles employed during experimentation. The robustness of the model's predictions is inherently linked to the quality and comprehensiveness of the underlying data.

The model is then used to make predictions on a validation dataset or new experimental data. It is worth noting that model validation can only be executed using an established database (thermo-kinetic parameters) and experimental inputs such as the recipe and temperature profiles. For instance, the work of Ghoreishi et al. [5] exemplifies the application of this modeling approach in the context of polyurethane gels. In their study, the authors meticulously predicted the thermo-kinetic parameters associated with various hydroxyl moieties to elucidate the impact of different hydroxyl groups on the reaction dynamics and temperature profiles of polyurethane systems [21]. To extract critical reactivity parameters, the researchers utilized urethane gel reaction temperature profiles derived from reference compounds, including 1-pentanol, 2-pentanol, and Voranol 360. This methodological framework not only facilitated a deeper understanding of the chemical behavior of polyurethane gels but also underscored the significance of hydroxyl group characteristics in influencing reaction kinetics.

3. **Feedback Loop**: A critical aspect of model evaluation involves systematically analyzing discrepancies between predicted and observed values. Such discrepancies can provide valuable insights into potential sources of error, which may arise from various factors, including the selection of input features, the architecture of the model, or the foundational assumptions made during the model development process. To effectively identify these sources of error, researchers typically thoroughly examine the input features utilized in the model. This includes assessing the relevance and accuracy of the data being fed into the model and ensuring that the features adequately represent the underlying physical processes being modeled. Additionally, the model architecture itself warrants scrutiny; this encompasses the choice of algorithms, the complexity of the model, and the appropriateness of the mathematical formulations employed. Furthermore, the underlying assumptions—such as the linearity of relationships, the independent variables, or the constancy of thermodynamic properties—must be critically evaluated, as any inaccuracies in these assumptions can significantly impact the model's predictive performance.

Upon identifying the sources of discrepancies, the model undergoes a refinement process, which is informed by the feedback obtained from the analysis. This refinement may involve several strategies, including the adjustment of hyperparameters, the incorporation of new features, or even a complete overhaul of the model architecture. For instance, if the initial model demonstrates an inability to adequately capture the effects of specific processing conditions—such as temperature, pressure, or humidity—on the thermal properties of RPUFs, researchers may introduce additional variables pertinent to those processing conditions. This could involve integrating real-time processing data or employing advanced feature engineering techniques to enhance the model's capacity to account for complex interactions. Moreover, iterative cycles of model evaluation and refinement are essential to developing robust predictive models. Each iteration allows for the continuous improvement of the model's accuracy and reliability, ultimately leading to a more nuanced understanding of the thermo-kinetic behavior of the materials under investigation. This iterative approach enhances the model's predictive capabilities and contributes to the broader field of materials science by providing insights that can inform experimental design and material optimization strategies.

4. **Result Database**: The models are stored in a result database once validated. The design of this database is crucial for facilitating efficient data retrieval and analysis. The database should be structured to allow for easy categorization and indexing of models based on various parameters, such as experimental conditions, model type, and performance metrics. Each stored model should be accompanied by comprehensive metadata detailing the experimental setup, data sources, model parameters, and validation results. This information is vital for future reference and researchers seeking to build upon existing work.

Hydroxyl Moiety Prediction. The hydroxyl groups in polyols are vital to the synthesis and performance of RPUFs. As bio-based polyols become more common, accurately characterizing their hydroxyl groups is crucial to enhancing foaming properties. However, this can be challenging, as Nuclear Magnetic Resonance (NMR) spectroscopy—the primary method for analyzing hydroxyl groups—is costly and inaccessible to many laboratories.

Dingcong et al. [49] proposed a computational approach to predict hydroxyl groups in bio-based polyols from lignocellulosic biomass. Their study utilized an iterative algorithm (illustrated in Fig. 2.3) and implemented it as a MATLAB script to focus on gelling reactions (Reactions 1–16 in Table 2.2), deemed sufficient for assessing the performance of each hydroxyl group. The script incorporated various functional codes: Bootstrap, Recipe, Database, Main, ReacSim, and FoamSim, each contributing to the accurate simulation of reaction kinetics (Fig. 2.6).

The script operates based on user-defined inputs, specifically the maximum temperature (T_{max}) and the designated time for the PU gelling reaction. The following six functions are integral to the script's operation: (1) The *Bootstrap function* is implemented to generate diverse combinations of polyol functionalities and hydroxyl groups. This is achieved through a for-loop process that systematically explores the foaming formulation parameters/values range. The function's output is the foundation for subsequent analyses by providing a comprehensive dataset of potential polyol properties. (2 and 3) The *Database and Recipe* functions manage the formulation data and estimate the relevant kinetics and thermochemical parameters. These parameters, detailed in Tables 2.3 and 2.5, include essential information such as the masses of polyols and isocyanates utilized in the formulation, specifically at an isocyanate index of 1.1. This structured approach ensures that all necessary data is readily accessible for further calculations. (4) The *Main function* is critical for validating the generated guesses regarding polyol properties. It employs MATLAB's ODE45 function to solve the ordinary differential equations (ODEs) defined in the ReacSim function. This rigorous validation ensures that only the most accurate estimates are retained for further analysis. (5) The *ReacSim function* encompasses the differential equations that model the instantaneous properties of the PU gelling reaction. These equations are essential for capturing the dynamic behavior of the reaction as it progresses over

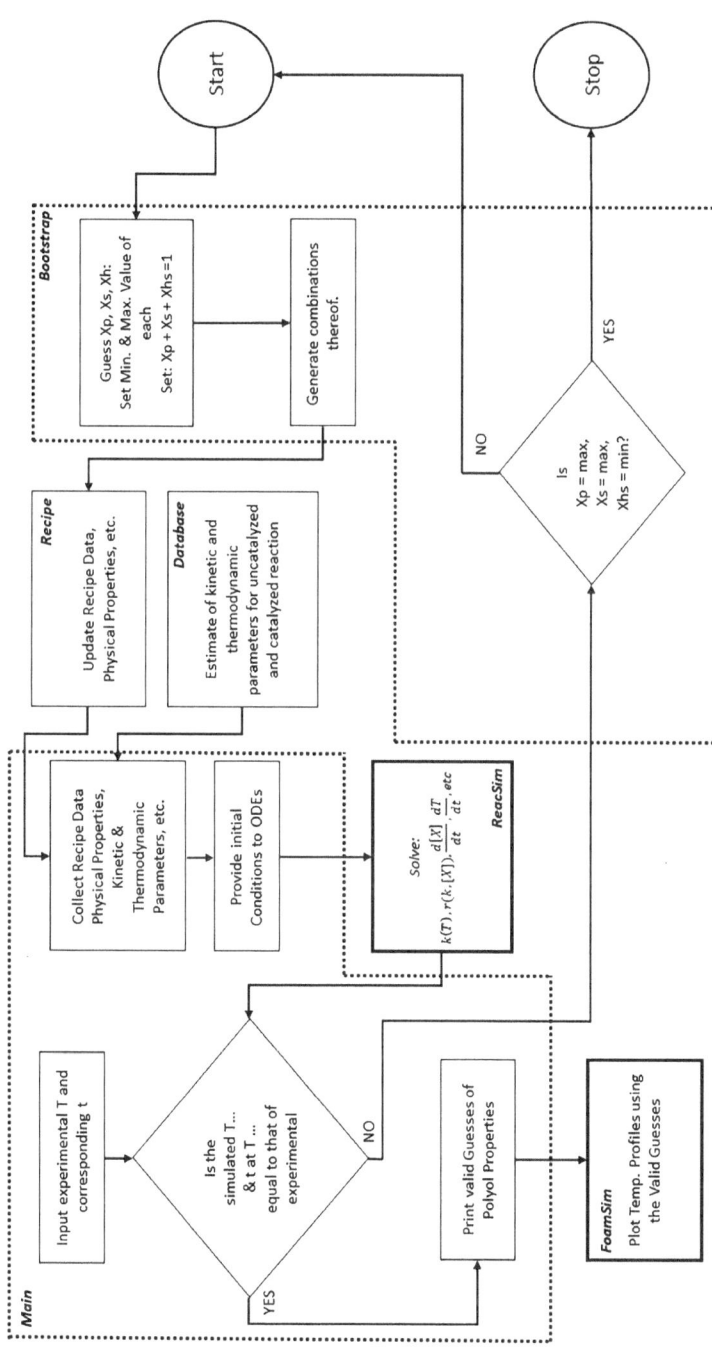

Fig. 2.6 MATLAB algorithm illustrating the looping process for simulating hydroxyl group fractions and their temperature profiles, utilizing six coded functions: Bootstrap, Recipe, Database, Main, ReacSim, and FoamSim. Reproduced from Dingcong et al. [49]. Under the CC BY 4.0 license

time, providing insights into the kinetics of the process. (6) Finally, the *FoamSim function* stores the validated guesses identified by the Main function. It also generates and displays the simulated temperature versus time (T vs. t) profiles, visually representing the gelling reaction dynamics. This function is crucial for interpreting the results and understanding the implications of the various parameter combinations.

Impact of Reaction Kinetics on the Rigid Polyurethane Foam Property. The conventional method for enhancing the properties of RPUFs typically involves modifying foam formulations through experimental techniques. However, this approach is often impeded by several practical challenges, such as the lack of access to specific experimental methods, high costs, complexity, time constraints, spatial limitations, waste generation, and associated risks. In response to these challenges, Alfeche et al. [21] introduced a thermo-kinetic in silico investigation approach to assess how varying reaction rates, represented by temperature profiles, affect the physico-mechanical properties of RPUFs. This study focused on the gelling reactions between coconut oil-based polyols and isocyanates.

The algorithm employed in this research is similar to that developed by Dingcong et al. [49]. It begins with the *UserInput* function, where essential data is gathered from the user, including formulation specifics, ingredient properties, and initial simulation conditions. Next, the *Bootstrap* function iteratively refines parameters to identify optimal fits. The *Database* is a repository for crucial kinetic and thermodynamic data, initial values, and assumptions that guide the simulation process. The *Main* function executes MATLAB's ODE45 solver on the *ReacSim* module, which contains differential equations modeling real-time property changes. The *MatchMatrices* function records valid parameter estimates and their corresponding numerical results throughout the simulation. Finally, *SimulatePlots* visualizes the validated simulations, presenting temperature profiles and gel times that elucidate the dynamics of the foam's reactions.

Notably, the script utilized datasets from prior studies, including thermo-kinetic parameters and methodologies. The thermo-kinetic model developed in this investigation demonstrated that reaction rates significantly influence the physicomechanical properties of the resulting RPUFs. For coconut oil-based polyols, a higher T_{max} is advantageous, as it facilitates the formation of more urethane bonds. Additionally, the duration required to reach T_{max} is critical, as it can substantially impact the foam's density, rheology, and morphology. Rapid gelling reactions may lead to ruptured cells due to premature cell expansion, which occurs when the developing polyurethane chains cannot withstand the expansion caused by blowing reactions. This finding aligns with observations from the experimental study conducted by Leng et al. [50]. Conversely, delayed gelling reactions produce foams with increased density and pseudo-ductile properties.

This study demonstrated that the hydroxyl groups in polyols critically influence gel reaction kinetics and, consequently, the resulting polyurethane network structure. Primary hydroxyl groups promote tighter crosslinking, leading to denser, mechanically robust PU networks, whereas secondary hydroxyl can form dangling chains that act as plasticizers, reducing rigidity. Hindered-secondary hydroxyls also limit crosslinking due to steric hindrance. Polyols rich in primary hydroxyl groups reach higher peak temperatures (T_{max}), yielding enhanced mechanical properties, as their reduced steric hindrance facilitates effective urethane formation, aligning with findings in the existing literature [51].

These findings enhanced the physico-mechanical properties of coconut oil-based RPUFs by employing computer simulations that focused on the reaction kinetics, particularly the gel time. The simulation methodology mitigated foam collapse and improved mechanical strength, notably reducing foam shrinkage and increased compressive strength. Maximizing T_{max} and regulating gel time through controlled catalyst loading, combined with the simulation approach to identify optimal foam formulations, proved effective. The simulation program also facilitated a deeper understanding of the results, as the simulated urethane concentration profiles corroborated the observed compressive strengths. Specifically, the coconut oil-based RPUF with a catalyst loading of 0.5 w/w% Cat 8 per bio-based polyol demonstrated a specific volume change and compressive strength that aligned closely with the hypothesized values, as indicated by the simulation results.

Exothermic Poly(Urethane-Urea) Rigid Polyurethane Foams. In pursuing material science advancement, the emergence of amine-hydroxyl hybrid polyols has gained significant attention in the polyurethane industries. These polyols commonly involve an amidation reaction during the functionalization process [50, 51]. The significant concentration of functional amine groups in these polyols is responsible for urea formations, making the produced foam a urethane-urea hybrid. Notably, urea bonds are known to have excellent mechanical properties, thus improving the RPUF's performance by a certain percentage. Building on this advancement, Fig. 2.7 presents a computational framework for optimizing reaction kinetics in RPUF synthesis, particularly relevant to amine-hydroxyl hybrid polyols. The iterative process integrates user-defined parameters, kinetic estimations, and thermodynamic modeling to refine the polymerization pathway. Given the strong urea formations in these polyols, the model systematically adjusts catalyst loading and polyol functionality to enhance mechanical performance. This computational approach ensures precise control over urethane-urea hybridization, leading to improved RPUF properties (Fig. 2.7).

The escalating demand for amine hybrid system polyols necessitates the development of efficient characterization methods to optimize their properties for RPUF applications. Significant advancements have been made in polyol synthesis, including tertiary amine-induced autocatalysis and the use of amino esters in equilibrium with diethanolamide, which enhance the mechanical and thermal properties of PU foams [52]. Analytical techniques that determine the fractional moieties of polyols provide critical insights into the resulting PU properties, as these moieties significantly influence the characteristics of PU networks as discussed previously. Additionally, amines enhance hydrogen bonding within the polymer matrix, thereby improving mechanical properties.

Despite these advancements, challenges remain in the characterization of hydroxyl functional moieties. Although standard quantitative methods exist for total hydroxyl numbers and primary and secondary amines, a comprehensive standard for quantifying specific fractions of primary, secondary, and hindered-secondary hydroxyls is lacking. This gap underscores the need for robust characterization methods for hydroxyl and amine moieties in traditional and hybrid polyols.

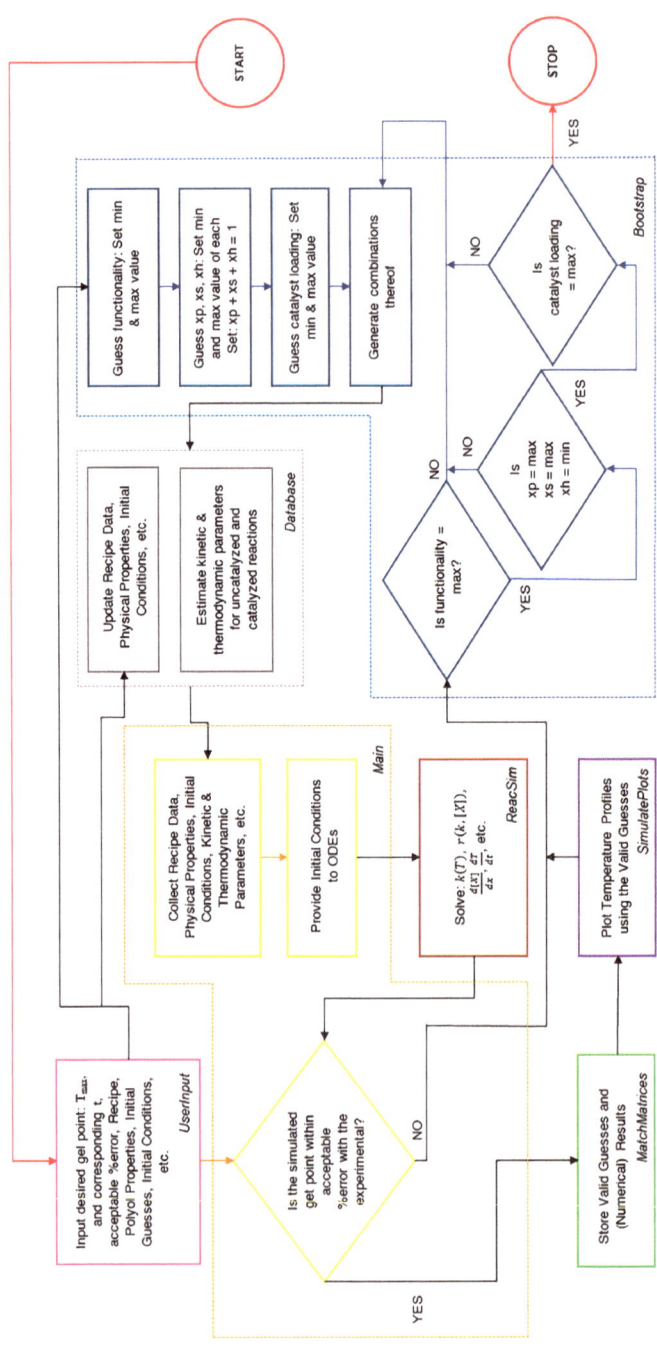

Fig. 2.7 Thermo-kinetic simulation algorithm used to evaluate the effects of reaction kinetics on the physico-mechanical properties of rigid polyurethane foam. Adapted from Alfeche et al. [21]. Under the CC BY 4.0 license

Moreover, understanding reaction kinetics is vital for tailoring material properties to meet specific application requirements. Recent simulations by Alfeche et al. [21] and Dingcong et al. [49] have illustrated that polymerization kinetics, particularly gel time, are significantly influenced by the chemical characteristics of polyols, including hydroxyl fractions. However, there remains a gap in identifying the thermo-kinetic properties associated with hybrid polyols, especially regarding their unique amine moieties and interactions with isocyanates.

Mendija et al. [1] addressed this gap by adopting a thermo-kinetic computational approach via MATLAB, simulating the polyurethane foaming temperature profiles based on hydroxyl and amine-functionalized polyols. Unlike the previously discussed models, the computational model in this work considers all possible reactions highlighted in Table 2.2. Generally, it considers both gelling and blowing reactions. In this case, blowing reactions are considered since this reaction usually generates intermediate primary amines, which ultimately react with isocyanate groups, forming urea bonds.

The algorithm used in this work considers six interrelated MATLAB function codes: User Input, Formulation, Database, Curve-Fitting, Prepolymer ODE, and Temperature Profile Generation (Fig. 2.8). This approach is similar to that in the literature [5, 21]. However, in this case, it considers additional functionalities, encompassing the amines functional groups and their impact on the RPUF's physico-mechanical properties. The *User Input* function is designed to collect essential data through four specific prompts: (1) the formulation of the isocyanate and polyol mixture, including the percentage of amine-based polyol substitution; (2) the property values of the amine-based polyol, which includes hydroxyl value, amine value, molecular weights, functionality, density, and specific heat capacities; (3) the property values of the isocyanate; and (4) experimental data, including reaction time and the range and intervals for the fractions of functional moieties. The hydroxyl value quantifies the concentration of hydroxyl groups, while the amine value indicates the levels of secondary amines present. Molecular weight is crucial for calculating molar quantities in prepolymerization reactions, and density is essential for determining the total volume of the reacting mixture, which influences reaction rates. Specific heat capacities are vital for evaluating the thermal properties of the chemical species involved, particularly for internal temperature calculations during polymerization. Additionally, the inputs for functional moiety fractions facilitate a thorough investigation into the combinations within the polyol system, ensuring accurate and reliable simulation outcomes.

The *Curve-Fitting* function determines the hydroxyl and amine moieties fractions by minimizing the average relative error to below 5.0% between experimental and simulated outcomes. This fitting process aligns experimental data with temperature–time profiles derived from ordinary differential equations (ODEs), minimizing the residuals between predicted and experimental data. The algorithm also accommodates the fitting of thermo-kinetic parameters or the fractions of polyol moieties, including primary, secondary, hindered-secondary hydroxyls, and secondary amines. The choice of fitting targets relies on the availability of existing data regarding these parameters.

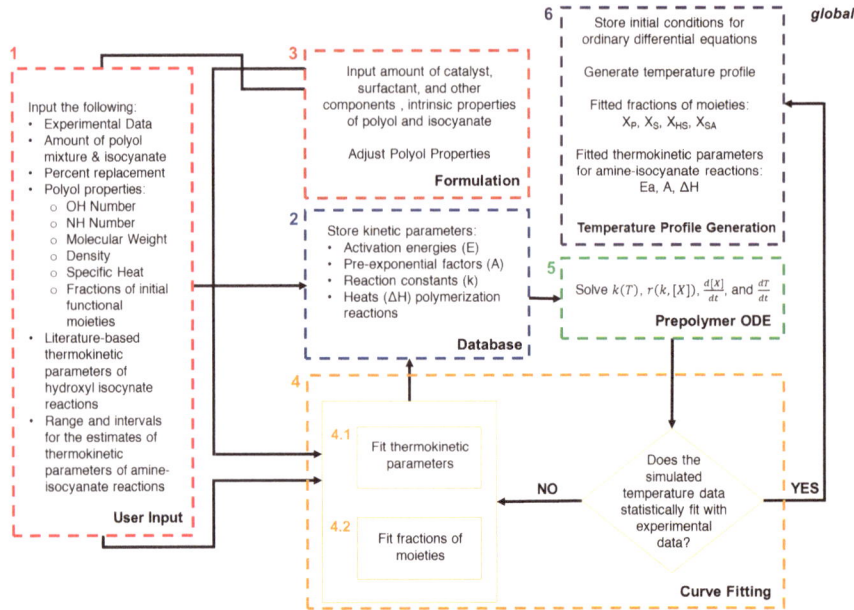

Fig. 2.8 Flowchart of the thermo-kinetic simulation algorithm used to model the temperature profile of polyurethane prepolymerization using hydroxyl-amine hybrid polyols. Adapted from Mendija et al. [1]. Under the CC BY 4.0 license

The *Temperature Profile Generation* function initializes the ordinary differential equations and constructs a temperature profile utilizing the MATLAB ode89 function. This function also identifies the gel point of polymerization, based on the premise that the temperature increase approaches zero. Consequently, the peak temperatures recorded during the prepolymerization of polyurethane foams are indicative of the gel point, marking the end of the polymerization reaction and the stabilization of temperature.

Collectively, the computational approach utilized in this study demonstrates significant potential for expediting the formulation and optimization of polyurethane hybrid foams through reaction process simulations. This method greatly reduces the necessity for extensive experimental trials, leading to substantial resource savings, cost reduction, and decreased exposure to hazardous solvents, thereby fostering a safer and more sustainable research environment. Moreover, as new polyol or isocyanate systems are developed, this methodology proves versatile in evaluating their reactivity and compatibility with other chemical components in polyurethane polymerization. By manipulating the type and concentration of functional groups within polyols, researchers can customize the reactivity and properties of polyurethane materials for targeted applications, particularly RPUFs, across various industries. Additionally, incorporating considerations of polymeric side reactions, such as the endothermic formation of allophanate and biuret segments in thermo-kinetic simulations, may enhance our understanding of the field and facilitate future advancements in polymer material development. This flexibility promotes sustainable innovation in polymer chemistry and engineering.

2.4   Traditional Versus Modern Methodology

Traditional methods, often based on empirical observations and simpler mathematical models, have historically helped characterize basic RPUF properties. However, advances in computational power and data analytics have given rise to modern techniques, including iterative methods, molecular dynamics (MD), and machine learning (ML) models, which offer greater predictive power and adaptability.

2.4.1   Methodology

Traditional. Traditional modeling techniques have historically relied on empirical observations and simpler mathematical models to characterize the basic properties of RPUF. One of the foundational methodologies involves using empirical and semi-empirical models to predict the mechanical properties of polyurethane foams. RPUF is frequently described using empirical models because of their ease of use and efficiency in connecting different physical characteristics with quantifiable metrics. Much research has been done on the relationship between the foam's density and compressive strength, leading to models correlating these variables. Research has shown that as the density of the foam increases, its compressive strength also tends to increase, which is a critical factor in applications requiring structural integrity [53]. Furthermore, to better understand how open-cell foams react to stress, micromechanical models have been suggested to characterize their behavior under large deformations [54]. Predictive models for the characteristics of RPUF have been created using mathematical modeling techniques, such as regression analysis. For example, Tu et al. established regression models to describe the effects of various formulations, such as the fraction of epoxidized soybean oil and the isocyanate index, on the compressive strength and thermal conductivity of water-blown rigid polyurethane foams [55]. These models help in optimizing formulations for desired performance characteristics. Unlike more intricate numerical techniques, empirical models are often simple and require less computing power. This enables academics and engineers lacking substantial knowledge of sophisticated modeling techniques to use them [56]. Empirical models can rapidly predict material properties based on easily measurable parameters, such as density and composition. This is particularly useful in the early stages of material development, where quick assessments are needed to guide formulation choices [57]. The tendency of empirical approaches to oversimplify intricate interactions between variables is one of their main drawbacks. Predictions may be inaccurate due to incomplete capture of the interactions between the various RPUF components, such as additives and environmental factors [58]. Empirical models have limited generalizability, are typically developed based on specific datasets, and may not generalize well to other formulations or processing conditions. For example, a model developed for a particular polyurethane foam type may not apply to another formulation with different chemical compositions or processing methods [59]. Cell size, form, and distribution are examples of RPUF's microstructural features that traditional empirical models may not sufficiently consider, even though these factors might substantially impact the material's overall characteristics. This error may result in erroneous mechanical and thermal behavior predictions of the RPUF [60].

In order to simulate and forecast the behavior of RPUF under varied conditions, researchers rely heavily on numerical approaches. Finite Element Analysis (FEA) and the Finite Volume Method (FVM) are popular numerical approaches in this context. Both methods provide valuable details regarding the mechanical, thermal, and dynamic properties of RPUF, which facilitates the optimization of foam compositions and applications. When modeling rigid polyurethane foam (RPUF), numerical techniques are essential because they allow researchers to simulate and forecast the material's behavior under many circumstances. Finite Element Analysis (FEA) and the Finite Volume Method (FVM) are popular numerical techniques in this context. Both approaches help to optimize foam formulations and applications by offering insightful information about the mechanical, thermal, and dynamic properties of RPUF. The mechanical behavior of materials, especially RPUF, is frequently analyzed using FEA, a potent numerical approach. FEA includes discretization, where foam structure is divided into a finite number of elements, creating a mesh that represents the geometry of the material. Each element is characterized by its material properties, which can vary across the foam [61]. The governing equations of motion or equilibrium are formulated for each element, typically using mechanics and material behavior principles. For RPUF, these equations often account for nonlinear material properties due to the foam's complex cellular structure [2]. The individual element equations are assembled into a global system representing the entire foam structure. In order to replicate real-world situations, boundary conditions and external loads are applied in this step [2].

Simulating the actual foaming process is a crucial component in modeling polyurethane foam. This includes simulating the dynamics of foam expansion and curing inside molds using numerical techniques like the FVM in mold-filling process modeling [62]. These models consider the physical properties of the blowing agents and chemical reactions producing foam. The considered processes are too complex to be visualized without using computational simulations of the interaction processes between RPUF and its environment [63] Similar to FEA, the domain is divided into discrete control volumes. However, in FVM, each control volume is subjected to conservation equations like mass, momentum, and energy rather than separate components [13]. The behavior of the foam is described by a set of algebraic equations obtained by integrating the governing equations over each control volume. This method works well when simulating the foaming process, where the interplay between the gas and liquid phases is crucial [64]. To ensure conservation principles are followed, the fluxes at the limits of each control volume are computed. Accurately describing the foaming process dynamics and the heat transfer properties of RPUF requires this step [65]. The repeated solutions of algebraic equations yield the solution fields in terms of temperature, pressure, and concentration, thereby providing information on thermal conductivity and the insulating properties of RPUF [13].

Although numerical techniques like FEA and FVM have many benefits when modeling RPUF, they also have drawbacks, including high computing costs, modeling assumptions, validation needs, and complicated foam structures. Complex geometries and high-resolution meshes can result in high computational demands, necessitating more sophisticated hardware and lengthier simulation periods [66]. Numerical model methods reduce assumptions on loading conditions, boundary conditions, and material responses. When the physical reality is not what exists in a model, such assumptions may lower the validity of the predicted values [60]. Numerical models always require validation against experimental data to ensure their validity. This process can be time-consuming and may require extensive experimental work to obtain the necessary data for validation [67]. The intricate cellular structure of RPUF can be challenging to represent accurately in numerical models. Simplifications made during the modeling process may overlook significant microstructural effects that influence the overall material behavior [65].

Modern. Modern RPUF modeling techniques enabled considerable progress by applying advanced computational methods, data analytics, and artificial intelligence. Improvements in model predictivity make models more flexible and, consequently, can more realistically simulate the behavior of RPUF under a wide variety of scenarios. Predictions related to databases allow easy access to large datasets; with material attributes and experimental outcomes, spotting patterns and connections becomes easier while speeding up the development of formulations [68]. Without conducting extensive experimental testing, researchers can develop well-informed predictions regarding the effectiveness of novel formulations by utilizing previous data [69]. Database models can help optimize formulations by identifying the best combinations of components based on previous successes, thus reducing the time and cost associated with trial-and-error approaches [70]. Statistical methods are employed to analyze the relationships between various parameters. Regression analysis is commonly used to develop predictive models that relate independent variables (e.g., formulation components) to dependent variables (e.g., mechanical properties). For example, Wang et al. studied the thermal decomposition kinetics of RPUF, providing insights that can be integrated into predictive models [71]. Predictive models are formulated using statistical analysis. A model can form linear or nonlinear equations according to the description given by the relationship of various properties. Creation of such models, in most cases, involves a machine learning approach to identifying intricate patterns from data that might not be addressed using other traditional methods [72]. By enabling researchers to use current data for formulation optimization, database predictions significantly reduce the time and resources needed for experimental testing [73]. However, the quality and completeness of the dataset have a significant impact on prediction accuracy. Predictions that are not accurate can result from incomplete or biased data [74].

A recent approach for modeling RPUF is iterative predictions, which entails improving predictive models by making repeated approximations in response to input from experimental data. This approach helps to optimize the features of RPUF for specific applications and capture its complicated behavior under different settings. The process begins with developing an initial predictive model based on empirical data, theoretical principles, or both. This model serves as the starting point for iterative refinement. For example, researchers may use kinetic models to simulate the reaction processes during foam formation, as seen in the work of Bernal et al., who discussed the kinetics of polyurethane foams [75]. The predictions made by the initial model are compared against the experimental data. Through this comparison, researchers can evaluate the correctness and dependability of the model by identifying differences between predicted and observed values [76]. By continuously improving models using real-world performance data, iterative prediction techniques enhance predictive accuracy over time [12]. Since these techniques may adjust to new data as they become available, they are appropriate for changing research settings where procedures and formulations are regularly modified [77]. For accurate results, the iterative procedure may take a long period and involve several testing and improvement iterations [78]. Implementing iterative predictions may require sophisticated statistical tools and expertise, which can be a barrier for some researchers [79].

Hybrid modeling often involves integrating empirical data with computational methods such as FEA or Computational Fluid Dynamics (CFD). For instance, empirical relationships derived from experimental data can inform the boundary conditions and material properties in FEA simulations, as demonstrated by Zhou et al. in their study on the weathering of roofing insulation materials [80]. Multiscale modeling, which combines several scales of analysis (such as molecular, microstructural, and macroscopic), can also be used in hybrid techniques. This makes it possible to comprehend how microstructural characteristics, like cell size and distribution, affect RPUF's macroscopic characteristics. However, the specific reference to Bernal et al. does not directly support this claim as it focuses on flexible polyurethane foams rather than RPUF [73]. Therefore, this citation has been removed. Recent advancements have seen the incorporation of machine learning algorithms into hybrid models. These algorithms can be implemented in datasets to trace complicated patterns and relationships that such conventional techniques tend to overlook. For instance, Uram et al. used machine learning to optimize a biochar-amended RPUF formulation for improved predictability [81]. Hybrid approaches offer a more comprehensive understanding of RPUF behavior, which combines various modeling techniques and captures microscopic and macroscopic processes [66]. Combined computational simulations and empirical data make predictions more accurate and reliable. This makes hybrid models especially helpful for complicated materials like RPUF [78]. Integrating multiple modeling techniques can lead to increased model development and interpretation complexity, requiring advanced knowledge and skills [82]. Hybrid modeling may demand more computational resources and time than simpler models, which can be a limiting factor in some research settings [83].

Techniques from machine learning (ML) and artificial intelligence (AI) have been widely implemented in the modeling of RPUF. Such methods can analyze large data sets and detect linkages and patterns overlooked by traditional modeling techniques. For instance, properties of RPUF may be predicted using machine learning algorithms trained by factors such as input density, formulation components, and processing conditions [84, 85]. AI-driven techniques can also automate the optimization process, enabling quick exploration of the formulation space and the discovery of optimal conditions for desired properties. In addition, utilizing real-time data from manufacturing processes can improve predictive modeling using AI-based approaches. For example, Holt et al. proposed a methodology to monitor the production processes of polymeric foams using methods inspired by monitoring techniques in thermo-kinetic-informed acoustic monitoring. This method results in real-time inputs, which are dynamically used to alter processing parameters [86]. Embedding AI with real-time monitoring is one significant step forward in the modeling and production of RPUF.

2.4.2 Costing

Cost Implications of Modern Techniques. Adopting modern modeling techniques requires a considerable initial investment in technology, including specialized software and high-performance computing resources. For instance, software licenses for FEM platforms or multiphysics simulation tools can be costly, and the computational hardware needed for large-scale simulations represents a substantial expense. However, these upfront costs are often offset by the long-term benefits that modern methodologies bring in terms of material efficiency, waste reduction, and improved product performance.

Traditional methods, while lower in immediate costs, often involve recurring expenses related to labor, materials, and time, particularly when repeated physical experiments are required. In contrast, modern simulations can reduce these ongoing expenses by minimizing the need for physical prototypes and enabling virtual testing of multiple formulation variables. As demonstrated in studies like that of Zhao et al. [2], advanced modeling can streamline research by identifying ideal material properties in silico, reducing the reliance on costly raw materials for trial-and-error testing.

Long-Term Financial Benefits. The predictive accuracy of modern techniques also yields long-term financial benefits by improving material performance and extending product life. For example, enhanced durability models, such as those developed by Yang et al. [87], enable researchers to design RPUFs with longer lifespans, reducing the frequency of replacement and lowering maintenance costs in applications such as building insulation. In industrial practice, these long-term improvements translate into substantial savings, as companies can offer products with superior performance and durability, reducing the total cost of ownership for customers and strengthening market position. In academic research, the cost considerations of modern modeling are also significant. Research institutions increasingly invest in computational resources to facilitate studies that contribute to foundational knowledge in materials science. The efficiencies gained through predictive modeling enhance the quality and scope of research outputs, making these investments worthwhile in terms of academic productivity and publication impact.

Decision-Making in Research and Industry. The cost of adopting modern modeling techniques is a key factor influencing decision-making in both research and industry. While traditional methods may be suitable for small-scale or exploratory projects, industries and academic institutions aiming for competitive advantage increasingly lean toward modern modeling. The efficiency gains, reduced material waste, and long-term cost savings make modern techniques a valuable investment, with a high return on investment that justifies the initial expenses.

References

1. L.C.C. Mendija et al., Elucidating the impact of polyol functional moieties on exothermic poly(urethane-urea) polymerization: a thermo-kinetic simulation approach. Sustainability **16**(11), 4587 (2024). https://doi.org/10.3390/su16114587
2. Y. Zhao, M.J. Gordon, A. Tekeei, F. Hsieh, G.J. Suppes, Modeling reaction kinetics of rigid polyurethane foaming process. J. Appl. Polym. Sci. **130**(2), 1131–1138 (2013). https://doi.org/10.1002/app.39287
3. R. Tesser, M. Di Serio, A. Sclafani, E. Santacesaria, Modeling of polyurethane foam formation. J. Appl. Polym. Sci. **92**(3), 1875–1886 (2004). https://doi.org/10.1002/app.20170
4. Y. Zhao, G.J. Suppes, Simulation of catalyzed urethane polymerization: an approach to expedite commercialization of bio-based materials. Catal. Surv. Asia **18**(2–3), 89–98 (2014). https://doi.org/10.1007/s10563-014-9168-9
5. R. Ghoreishi, Y. Zhao, G.J. Suppes, Reaction modeling of urethane polyols using fraction primary secondary and hindered-secondary hydroxyl content. J. Appl. Polym. Sci. **131**(12), 40388 (2014). https://doi.org/10.1002/app.40388
6. H. Al-Moameri, R. Ghoreishi, G. Suppes, Impact of inter- and intra-molecular movements on thermoset polymerization reactions. Chem. Eng. Sci. **161**, 14–23 (2017). https://doi.org/10.1016/j.ces.2016.12.007
7. H. Al-Moameri, L. Jaf, G.J. Suppes, Viscosity-dependent frequency factor for modeling polymerization kinetics. RSC Adv. **7**(43), 26583–26592 (2017). https://doi.org/10.1039/C7RA01242J
8. H. Al-Moameri, L. Jaf, G.J. Suppes, Simulation approaches for the mechanisms of thermoset polymerization reactions. Mol. Catal. **504**, 111485 (2021). https://doi.org/10.1016/j.mcat.2021.111485
9. W.H. Carothers, Polymerization. Chem. Rev. **8**(3), 353–426 (1931). https://doi.org/10.1021/cr60031a001
10. W.H. Carothers, Polymers and polyfunctionality. Trans. Faraday Soc. **32**, 39 (1936). https://doi.org/10.1039/tf9363200039
11. S.A. Baser, D.V. Khakhar, Modeling of the dynamics of water and R-11 blown polyurethane foam formation. Polym. Eng. Sci. **34**(8), 642–649 (1994). https://doi.org/10.1002/pen.760340805
12. H. Al-Moameri, Y. Zhao, R. Ghoreishi, G.J. Suppes, Simulation blowing agent performance, cell morphology, and cell pressure in rigid polyurethane foams. Ind. Eng. Chem. Res. **55**(8), 2336–2344 (2016). https://doi.org/10.1021/acs.iecr.5b04711

13. H. Abdessalam, B. Abbès, Y.M. Li, Y.Q. Guo, E. Kwassi, J.L. Romain, Polyurethane foaming process modeling by finite point method. Adv. Mater. Res. **881–883**, 841–845 (2014)
14. Y.B. Kim, Numerical simulation of PU foaming flow in a refrigerator cabinet. J. Cell. Plast. **41**, 251–266 (2005)
15. C. Raimbault, Foaming parameter identification of polyurethane using FOAMAT® device. Polym. Eng. Sci. **61**, 1243–1256 (2021)
16. O. Weißenborn, C. Ebert, M. Gude, Modeling of the strain rate dependent deformation behaviour of rigid polyurethane foams. Polym. Testing **54**, 145–149 (2016)
17. K. Valli Priyadharshini, A. Vijay, K. Swaminathan, T. Avudaiappan, V. Banupriya, Materials property prediction using feature selection based machine learning technique. Mater. Today Proc. **69**, 710–715 (2022). https://doi.org/10.1016/j.matpr.2022.07.134
18. H. Sun, H. Zhang, G. Ren, C. Zhang, A knowledge transfer framework for general alloy materials properties prediction. Materials **15**(21), 7442 (2022). https://doi.org/10.3390/ma15217442
19. W. Ma, C. Hu, S. Guo, Z. Zhao, T. Huang, Flexural and shear bond performance of polyurethane-mortar interface under micro- and macroscale. J. Mater. Civ. Eng. **31**(7), 04019105 (2019). https://doi.org/10.1061/(ASCE)MT.1943-5533.0002728
20. R. Hammad, S. Mondal, Predicting poisson's ratio: a study of semisupervised anomaly detection and supervised approaches. ACS Omega **9**(1), 1956–1961 (2024). https://doi.org/10.1021/acsomega.3c08861
21. F.L.A.M. Alfeche et al., In silico investigation of the impact of reaction kinetics on the physico-mechanical properties of coconut-oil-based rigid polyurethane foam. Sustainability **15**(9), 7148 (2023). https://doi.org/10.3390/su15097148
22. K.T. Butler, D.W. Davies, H. Cartwright, O. Isayev, A. Walsh, Machine learning for molecular and materials science. Nature **559**(7715), 547–555 (2018). https://doi.org/10.1038/s41586-018-0337-2
23. C.R. Groom, I.J. Bruno, M.P. Lightfoot, S.C. Ward, The cambridge structural database. Acta Crystallogr. Sect. B Struct. Sci. Cryst. Eng. Mater. **72**(2), 171–179 (2016). https://doi.org/10.1107/S2052520616003954
24. S. Gražulis et al., Crystallography open database (COD): an open-access collection of crystal structures and platform for world-wide collaboration. Nucleic Acids Res. **40**(D1), D420–D427 (2012). https://doi.org/10.1093/nar/gkr900
25. D. Torelli, H. Moustafa, K.W. Jacobsen, T. Olsen, High-throughput computational screening for two-dimensional magnetic materials based on experimental databases of three-dimensional compounds. NPJ Comput. Mater. **6**(1), 158 (2020). https://doi.org/10.1038/s41524-020-00428-x
26. A. Merkys, A. Vaitkus, A. Grybauskas, A. Konovalovas, M. Quirós, S. Gražulis, Graph isomorphism-based algorithm for cross-checking chemical and crystallographic descriptions. J. Cheminformatics **15**(1), 25 (2023). https://doi.org/10.1186/s13321-023-00692-1
27. I. Bruno, S. Gražulis, J.R. Helliwell, S.N. Kabekkodu, B. McMahon, J. Westbrook, Crystallography and databases. Data Sci. J. **16**, 38 (2017). https://doi.org/10.5334/dsj-2017-038
28. A. Zakutayev et al., An open experimental database for exploring inorganic materials. Sci. Data **5**(1), 180053 (2018). https://doi.org/10.1038/sdata.2018.53
29. L. Talirz et al., Materials cloud, a platform for open computational science. Sci. Data **7**(1), 299 (2020). https://doi.org/10.1038/s41597-020-00637-5
30. M. Aykol et al., Network analysis of synthesizable materials discovery. Nat. Commun. **10**(1), 2018 (2019). https://doi.org/10.1038/s41467-019-10030-5
31. C.W. Andersen et al., OPTIMADE, an API for exchanging materials data (2021). https://doi.org/10.48550/ARXIV.2103.02068
32. M. Dreger, K. Malek, M.J. Eslamibidgoli, M.H. Eikerling, Synergizing ontologies and graph databases for highly flexible materials-to-device workflow representations Chemistry (2023). https://doi.org/10.26434/chemrxiv-2023-f3bc5
33. Z. Sa'adi et al., Comparative assessment of empirical random forest family's model in simulating future streamflow in different basin of Sarawak, Malaysia. J. Atmospheric Sol.-Terr. Phys. **265**, 106381 (2024). https://doi.org/10.1016/j.jastp.2024.106381

34. A. Holzinger, Interactive machine learning for health informatics: when do we need the human-in-the-loop? Brain Inform. **3**(2), 119–131 (2016). https://doi.org/10.1007/s40708-016-0042-6
35. K.T. Schütt, H.E. Sauceda, P.-J. Kindermans, A. Tkatchenko, K.-R. Müller, SchNet—a deep learning architecture for molecules and materials. J. Chem. Phys. **148**(24), 241722 (2018). https://doi.org/10.1063/1.5019779
36. M.T. Cretu, A. Toniato, A.C. Vaucher, A. Thakkar, A. Debabeche, T. Laino, Standardization of chemical compounds using language modeling. Chemistry (2022). https://doi.org/10.26434/chemrxiv-2022-14ztf
37. M. Haghighatlari, J. Li, F. Heidar-Zadeh, Y. Liu, X. Guan, T. Head-Gordon, Learning to make chemical predictions: the interplay of feature representation, data, and machine learning algorithms (2020). https://doi.org/10.48550/ARXIV.2003.00157
38. F. Wan, J. Michael Zeng, Deep learning with feature embedding for compound-protein interaction prediction. Bioinformatics (2016). https://doi.org/10.1101/086033
39. S. Shermukhamedov, D. Mamurjonova, T. Maihom, M. Probst, Structure to property: machine learning methods for predicting electronic properties of crystals. Chemistry (2023). https://doi.org/10.26434/chemrxiv-2023-gq77g
40. Z. Han, S. Yin, Research on Semi-supervised Classification with an Ensemble Strategy. in *Proceedings of the 2016 4th International Conference on Sensors, Mechatronics and Automation (ICSMA 2016)* (Atlantis Press, Zhuhai, China, 2016). https://doi.org/10.2991/icsma-16.2016.119
41. T. Provoost, M.-F. Moens, Semi-supervised Learning for the BioNLP gene regulation network. BMC Bioinformatics **16**(S10), S4 (2015). https://doi.org/10.1186/1471-2105-16-S10-S4
42. E.A. Olivetti et al., Data-driven materials research enabled by natural language processing and information extraction. Appl. Phys. Rev. **7**(4), 041317 (2020). https://doi.org/10.1063/5.0021106
43. A.L. Ferguson, Machine learning and data science in soft materials engineering. J. Phys. Condens. Matter **30**(4), 043002 (2018). https://doi.org/10.1088/1361-648X/aa98bd
44. A. Glielmo, B.E. Husic, A. Rodriguez, C. Clementi, F. Noé, A. Laio, Unsupervised learning methods for molecular simulation data. Chem. Rev. **121**(16), 9722–9758 (2021). https://doi.org/10.1021/acs.chemrev.0c01195
45. J. Wang, Iterative pseudo-labelling with SoftMax probability in text classification. Appl. Comput. Eng. **6**(1), 24–29 (2023). https://doi.org/10.54254/2755-2721/6/20230738
46. C.M. Eckhardt et al., Unsupervised machine learning methods and emerging applications in healthcare. Knee Surg. Sports Traumatol. Arthrosc. **31**(2), 376–381 (2023). https://doi.org/10.1007/s00167-022-07233-7
47. H. Huo et al., Semi-supervised machine-learning classification of materials synthesis procedures. Npj Comput. Mater. **5**(1), 62 (2019). https://doi.org/10.1038/s41524-019-0204-1
48. C. Kunselman, V. Attari, L. McClenny, U. Braga-Neto, R. Arroyave, Semi-supervised learning approaches to class assignment in ambiguous microstructures. Acta Mater. **188**, 49–62 (2020). https://doi.org/10.1016/j.actamat.2020.01.046
49. R.G. Dingcong et al., An iterative method for the simulation of rice straw-based polyol hydroxyl moieties. Sustainability **15**(15), 12082 (2023). https://doi.org/10.3390/su151512082
50. X. Leng et al., A study on coconut fatty acid diethanolamide-based polyurethane foams. RSC Adv. **12**(21), 13548–13556 (2022). https://doi.org/10.1039/D2RA01361D
51. L.N.A. Hipulan et al., Development of high-performance coconut oil-based rigid polyurethane-urea foam: a novel sequential amidation and prepolymerization process. ACS Omega, 3c09598 (2024). https://doi.org/10.1021/acsomega.3c09598
52. R.G. Dingcong et al., A novel reaction mechanism for the synthesis of coconut oil-derived biopolyol for rigid poly(urethane-urea) hybrid foam application. RSC Adv. **13**(3), 1985–1994 (2023). https://doi.org/10.1039/D2RA06776E
53. F. Saint-Michel, L. Chazeau, J. Cavaillé, E. Chabert, Mechanical properties of high density polyurethane foams: I. effect of the density. Compos. Sci. Technol. **66**(15), 2700–2708 (2006)
54. L. Marşavina, T. Sadowski, D.M. Constantinescu, R. Negru, Polyurethane foams behaviour. Experiments versus modeling. Key Eng. Mater. **399**, 123–130 (2008)

55. Y. Tu, H. Fan, G.J. Suppes, F. Hsieh, Physical properties of water-blown rigid polyurethane foams containing epoxidized soybean oil in different isocyanate indices. J. Appl. Polym. Sci. **114**(5), 2577–2583 (2009)
56. M. Günther, A. Lorenzetti, B. Schartel, Fire phenomena of rigid polyurethane foams. Polymers **10**(10) (2018)
57. M.T. Hoang, C. Perrot, Identifying local characteristic lengths governing sound wave properties in solid foams. J. Appl. Phys. **113**(8) (2013)
58. R. Atiénzar-Navarro, R.d.R. Tormos, J.A. Fernández, V.J. Sánchez-Morcillo, R. Picó, Sound absorption properties of perforated recycled polyurethane foams reinforced with woven fabric. Polymers **12**(2), 401 (2020)
59. Y. He, D. Qiu, Z. Yu, Study of failure behaviors of rigid polyurethane foam treated under thermal and vibration conditions by experiment and numerical simulation. J. Appl. Polym. Sci. **140**(4) (2022)
60. R. Ippili, P. Davies, A. Bajaj, L. Hagenmeyer, Nonlinear multi-body dynamic modeling of seat–occupant system with polyurethane seat and h-point prediction. Int. J. Ind. Ergon. **38**(5–6), 368–383 (2008)
61. H. Andami, Performance assessment of rigid polyurethane foam core sandwich panels under blast loading. Int. J. Prot. Struct. **11**(1), 109–130 (2019)
62. S. Geier, C. Winkler, M. Piesche, Numerical simulation of mold filling processes with polyurethane foams. Chem. Eng. Technol. **32**(9), 1438–1447 (2009)
63. L. Yeon, J. Nam, J. Ryu, Numerical analysis on foam reaction injection molding of polyurethane, part b: parametric study and real application. J. Korean Crystal Growth Crystal Technol. **26**(6), 258–262 (2016)
64. D. Qiu, Y. He, Z. Yu, Investigation on compression mechanical properties of rigid polyurethane foam treated under random vibration condition: an experimental and numerical simulation study. Materials **12**(20), 3385 (2019)
65. J. Wu, Y. He, Z. Yu, Failure mechanism of rigid polyurethane foam under high temperature vibration condition by experimental and finite element method. J. Appl. Polym. Sci. **137**(6) (2019)
66. M. Zieleniewska et al., Preparation and characterisation of rigid polyurethane foams using a rapeseed oil-based polyol. Ind. Crops Prod. **74**, 887–897 (2015)
67. S. Polimera, A. Gali, S.K. Nath, A. Rahaman, M.R. Chandan, S.J. Balakumaran, Thermo-mechanical property enhancement of rigid polyurethane foam using silica and alumina as hybrid fillers over single filler. Polym. Compos. **44**(10), 6454–6466 (2023)
68. M. Leszczyńka et al., Vegetable fillers and rapeseed oil-based polyol as natural raw materials for the production of rigid polyurethane foams. Materials **14**(7) (2021)
69. M. Akkoyun, Ş. Akkoyun, Blast furnace slag or fly ash filled rigid polyurethane composite foams: a comprehensive investigation. J. Appl. Polym. Sci. **136**(20) (2019)
70. M. Akkoyun, E. Suvacı, Effects of TIO_2, ZNO, and FE_3O_4 nanofillers on rheological behavior, microstructure, and reaction kinetics of rigid polyurethane foams. J. Appl. Polym. Sci. **133**(28) (2016)
71. S. Wang, H. Chen, L. Zhang, Thermal decomposition kinetics of rigid polyurethane foam and ignition risk by a hot particle. J. Appl. Polym. Sci. **131**(4) (2013)
72. V.M. Gravit, O. Ogidan, E. Znamenskaya, Methods for determining the number of closed cells in rigid sprayed polyurethane foam. MATEC Web Conf. **193**, 03027 (2018)
73. F. Wang, J. Liang, Q. Tang, Preparation and properties of rigid polyurethane foams reinforced by sepiolite minerals nanofibers. Key Eng. Mater. **512–515**, 280–283 (2012)
74. A.M. Raji, H. Hambali, Z.I. Khan, Z. Mohamad, A. Hassan, R. Ogabi, Emerging trends in flame retardancy of rigid polyurethane foam and its composites: a review. J. Cell. Plast. **59**(1), 65–122 (2022)
75. M.M. Bernal et al., Effect of hard segment content and carbon-based nanostructures on the kinetics of flexible polyurethane nanocomposite foams. Polymer **53**(19), 4025–4032 (2012)
76. Y. Fan, A. Gomez, A. Muliana, V.L. Saponara, Multi-scale analysis of diffusion of fluid in sandwich composites. Polym. Compos. **40**(9), 3520–3532 (2019)

77. T. Zhu, S. Chen, W. Zhu, Y. Wang, Optimization of sound absorption property for polyurethane foam using adaptive simulated annealing algorithm. J. Appl. Polym. Sci. **135**(26) (2018)
78. S. Liu, H. Luo, K. Xv, W. Qiu, P. Chen, Preparation and characterization of gf modified waste rigid polyurethane foam. Materiale Plastice **57**(4), 275–285 (2021)
79. C. Zhang et al., The foaming dynamic characteristics of polyurethane foam. J. Cell. Plast. **56**(3), 279–295 (2019)
80. S. Zhou et al., Weathering of roofing insulation materials under multi-field coupling conditions. Materials **12**(20), 3348 (2019)
81. S. Polimera, A. Gali, A. Rahaman, M.R. Chandan, S.J. Balakumaran, S.K. Nath, Thermo-mechanical property enhancement of rigid polyurethane foam composite using low cost, environment friendly fly ash over silica particles. J. Vinyl Add. Tech. **30**(1), 156–171 (2023)
82. K. Uram, M. Kurańska, J. Andrzejewski, A. Prociak, Rigid polyurethane foams modified with biochar. Materials **14**(19), 5616 (2021)
83. S. Członka, A. Kairytė, K. Miedzińska, A. Strąkowska, Casein/apricot filler in the production of flame-retardant polyurethane composites. Materials **14**(13), 3620 (2021)
84. S. Michałowski, K. Pielichowski, 1,2-propanediolizobutyl poss as a co-flame retardant for rigid polyurethane foams. J. Therm. Anal. Calorim. **134**(2), 1351–1358 (2018)
85. J.P. Soares Kaiser, J.M.d. Silva Neto, H.L. Corrêa, Analysis of the effects of region and direction on the mechanical strength of polyurethane foams in white goods refrigerators. Revista Produção Online **23**(1), 4953 (2023)
86. J. Holt, C. Torres-Sánchez, P. Conway, Monitoring the continuous manufacture of a polymeric foam via a thermokinetic-informed acoustic technique. Proc. Inst. Mech. Eng., Part E: J. Process Mech. Eng. **235**(6), 1998–2007 (2021)
87. C. Yang, L. Fischer, S. Maranda, J. Worlitschek, Rigid polyurethane foams incorporated with phase change materials: a state-of-the-art review and future research pathways. Energy Build. **105**, 164–178 (2015)

Chapter 3
Key Factors in Computational Modeling of RPUF

3.1 Prediction Method and Modeling Technique

In the development of rigid polyurethane foam (RPUF), predictive modeling is an indispensable tool, allowing researchers to anticipate material behavior under various conditions and optimize properties for a range of applications. RPUFs are known for their high thermal insulation, low density, and mechanical stability, making them valuable in industries such as automotive, construction, and aerospace. However, the production of RPUFs involves a complex, exothermic foaming process, which includes chemical reactions and the expansion of cells driven by physical blowing agents (PBAs). Each factor involved in this process—chemical kinetics, heat transfer, and cellular structure—contributes to the final properties of RPUF, such as thermal conductivity, strength, and flammability.

Predictive modeling techniques, encompassing reaction kinetics, thermomechanical simulations, and multiscale modeling, offer deep insights into these complex processes, enabling researchers to fine-tune RPUF formulations and process conditions effectively. These methods simulate the foam's behavior under various stressors, allowing researchers to optimize properties like mechanical strength, thermal insulation, durability, and fire resistance. Finite Element Modeling (FEM) is widely used as part of these predictive efforts, providing a robust framework for simulating physical responses in RPUF and integrating multiple modeling techniques for comprehensive analysis. This section explores the objectives of predictive modeling in RPUF research, examining examples from recent studies and discussing applications across various modeling techniques.

A. A. Lubguban et al., *Computational Thermo-kinetics of Rigid Polyurethane Foams*, SpringerBriefs in Applied Sciences and Technology, https://doi.org/10.1007/978-981-96-2077-7_3

3.1.1 Objectives of Prediction Methods in RPUF Research

Predictive modeling in RPUF serves several primary objectives, each focused on enhancing a specific material property to meet performance standards across various applications. The most common objectives include optimizing mechanical strength, enhancing thermal insulation, improving long-term durability, and augmenting fire resistance. Each objective targets a critical aspect of RPUF's functionality, with predictive modeling techniques providing the necessary tools to evaluate, modify, and enhance these properties.

Optimizing Mechanical Strength. Mechanical strength is crucial for RPUFs used in load-bearing applications, such as automotive and aerospace components. The foam's internal cellular structure—a network of interconnected open and closed cells—plays a significant role in determining its compressive and tensile properties. Factors such as cell size, distribution, density, and crosslinking density within the polymer matrix are essential parameters in RPUF, affecting its ability to withstand mechanical stress without deformation. Predictive models simulate the stress–strain behavior of RPUF, enabling researchers to modify and optimize these parameters for improved structural integrity.

Dingcong et al. [1] focused on examining the mechanical performance of RPUFs under compressive loads. Their study used simulation models to analyze the relationships between formulation variables and compressive strength, which provided insights into optimizing foam composition for applications requiring high compressive resistance, such as in automotive interiors and building insulation. By mapping the stress distribution and strain characteristics across different foam densities, the study illustrated how variations in foam formulation impact mechanical stability, contributing to the creation of RPUFs that balance compressive strength with lightweight structure. Alfeche et al. [2] further demonstrated the role of predictive modeling in enhancing RPUF's mechanical properties, specifically in the context of bio-based foams. Their research explored coconut oil-based RPUFs, leveraging a combination of FEM and reaction kinetics modeling to analyze crosslinking density and its impact on foam stiffness and durability. The study found that higher crosslink densities result in improved compressive strength, enhancing the foam's rigidity and load-bearing capabilities while maintaining flexibility. Such findings underscore the value of predictive models in achieving sustainable, high-performance RPUFs that align with environmental goals and offer mechanical robustness. In high-impact applications, where materials are exposed to dynamic loads, predictive modeling provides insights into the foam's response under specific loading rates. For example, Whisler and Kim [3] examined high-strain dynamic loading scenarios to understand strain-rate sensitivity in RPUF. Through simulated uniaxial and hydrostatic compression tests, they discovered that RPUF exhibits varying mechanical responses depending on the strain rate, with higher loading rates resulting in increased modulus and higher stress at collapse. This strain-rate sensitivity is particularly relevant for applications in automotive crash-resistant structures, where predictive modeling helps in selecting RPUF formulations optimized for high-speed impacts.

Enhancing Thermal Insulation. Thermal insulation is one of the primary applications of RPUF, especially in industries where temperature regulation and energy efficiency are paramount, such as building insulation, refrigeration, and thermal management. RPUF's thermal insulating properties are largely attributed to its closed-cell structure, which traps gases within the cells, reducing heat transfer through conduction. PBAs play an essential role in the foaming process, influencing the cell formation, distribution, and, consequently, the thermal conductivity of the foam. Predictive models focused on heat transfer dynamics and PBA behavior are indispensable for maximizing RPUF's insulation performance, as they simulate how different variables affect internal temperature, thermal conductivity, and cellular integrity.

Wang et al. [4] investigated the role of PBAs in RPUF's thermal insulation capabilities, using thermomechanical simulations to explore how different PBA types impact internal temperature and foam stability during curing. Their predictive model revealed that the efficiency of PBAs directly affects cell structure and heat resistance, with potential applications in RPUFs designed for stringent insulation requirements, such as those in refrigeration and construction. The study found that PBAs with higher heat absorption efficiency create more stable cell structures, resulting in foams with reduced thermal conductivity and enhanced insulation.

Phase change materials (PCMs) have also been explored as an additive in RPUF to enhance its thermal storage capacity. Yang et al. (2015) reviewed the incorporation of PCMs in RPUF, specifically focusing on the role of PCM content in heat absorption and thermal stability. Their multiscale predictive model simulated PCM behavior within the foam matrix, demonstrating that controlled distribution of PCM particles significantly improves thermal insulation without compromising mechanical strength. By simulating heat flow within RPUF with PCM additives, this model provided essential insights into the optimal PCM concentration for high-efficiency insulation, making RPUFs suitable for energy-efficient building materials where both insulation and thermal storage are critical [5].

Improving Long-Term Durability. Durability is a crucial attribute for RPUFs exposed to harsh environmental conditions, such as temperature fluctuations, humidity, and UV exposure. These factors can degrade RPUF's mechanical and thermal properties over time, leading to reduced performance and material failure. Predictive models that simulate these environmental impacts help researchers anticipate degradation rates and guide the development of formulations with extended lifespans, ensuring that RPUF meets durability standards required for applications in automotive, outdoor, and industrial environments. Al-Moameri et al. [5] applied predictive modeling to study the thermal and oxidative degradation of bio-based RPUFs, which often exhibit different degradation profiles compared to petroleum-based foams. Their model integrated reaction kinetics and thermomechanical properties to assess long-term stability by simulating the breakdown of chemical bonds under thermal stress. By identifying optimal polyol types and crosslink densities, the model provided a pathway for designing RPUFs with enhanced durability, a property critical in automotive and outdoor applications where foams must withstand prolonged exposure to environmental elements [6]. Yang et al. [6] investigated the durability of phase-change-based RPUFs under variable environmental conditions. By simulating the interaction between PCM particles and the foam matrix, their model evaluated the stability of foam properties over multiple thermal cycles, showing how PCM integration can impact RPUF's mechanical stability and insulation performance in extreme temperature environments. Such durability-focused predictive models are indispensable for applications where RPUF must maintain its structural integrity and insulating capabilities despite environmental fluctuations [5].

Enhancing Fire Resistance. RPUFs are inherently flammable, which limits their application in sectors with strict fire safety requirements, such as building and transportation. To address this, flame-retardant additives are incorporated into RPUF formulations to enhance fire resistance. Predictive models that simulate thermal degradation, char formation, and combustion behavior provide a framework for evaluating the performance of flame retardants, enabling researchers to identify optimal concentrations and compositions that improve fire resistance without negatively affecting foam properties. Usta [7] conducted a study on RPUF modified with intumescent flame retardants, using predictive modeling to evaluate the impact of different formulations on fire behavior. The model simulated the thermal degradation and heat release rates of RPUFs with varying levels of flame retardants, revealing that specific additives form protective char layers that limit combustion. Such models are essential in developing fire-resistant RPUFs that meet safety standards in building and transportation applications, where compliance with fire resistance regulations is crucial.

3.1.2   Selection of Modeling Techniques Based on Research Objectives

The choice of modeling technique is integral to achieving specific objectives in RPUF research. The complexity of RPUF's foaming process and its dependence on multiple interacting variables necessitate the use of a variety of modeling approaches. Each technique offers unique advantages and limitations, and the optimal selection often depends on the desired outcome, whether it's to analyze reaction kinetics, simulate heat transfer, or assess structural integrity. The following sections discuss key modeling techniques—reaction kinetics models, thermomechanical simulations, FEM, and multiscale models—and their applications in RPUF research.

Reaction Kinetics Models. Reaction kinetics models are essential for predicting the exothermic reactions driving RPUF foaming, providing insights into reaction rates, gel time, and polymerization. By solving differential equations that represent reaction kinetics, these models allow researchers to control and optimize the curing process. Zhao et al. [8] developed a MATLAB-based model that simulated temperature profiles during the foaming process, enabling precise control over exothermic reactions, particularly in bio-based polyol systems. This approach is crucial in managing heat generation and ensuring a uniform cell structure, particularly in RPUFs intended for thermal insulation.

Thermomechanical Simulations. Thermomechanical simulations assess how RPUF responds to simultaneous thermal and mechanical stresses, making them useful in applications where foam stability under thermal gradients is essential. These models simulate how variations in temperature impact stress distribution within the foam, helping researchers evaluate the foam's behavior in environments like automotive interiors or structural insulation. Permann et al. [9] utilized a multiphysics framework to combine thermal and mechanical simulations, providing insights into RPUF's behavior under high-temperature conditions and supporting the development of foams with enhanced stability.

Multiscale Models. Multiscale models, which bridge molecular-level interactions with macroscopic properties, are critical in applications that demand precise control over microstructural properties. Yang et al. employed a multiscale approach to simulate the integration of PCMs within RPUF, examining how molecular interactions influence bulk thermal conductivity and mechanical performance. Multiscale modeling enables accurate predictions of material behavior, making it invaluable in optimizing additives like PCMs for energy-efficient applications [5].

3.1.3 Application of Prediction Methods in RPUF Research

Predictive modeling has revolutionized RPUF research, facilitating advancements in foam properties across mechanical, thermal, and environmental domains.

Mechanical Simulations for Structural Applications. Predictive models focused on mechanical strength have led to significant improvements in RPUF for load-bearing applications. Dingcong et al. [1] used mechanical simulations to optimize compressive strength in RPUFs with various formulation adjustments, helping create lightweight yet robust foams for automotive and aerospace components. This approach exemplifies the potential of predictive models to drive innovations in structural applications.

Thermal and Fire Resistance Modeling. Models focused on thermal and fire resistance are essential in designing safe and durable RPUFs for building and transportation. Usta's [7] predictive model for flame-retardant RPUFs demonstrated how simulations of thermal degradation guide the selection of effective fire-retardant additives, supporting the development of foams that comply with fire safety regulations.

Environmental Durability Predictions. As sustainability becomes a priority, predictive models that simulate long-term environmental impact help develop durable RPUFs that withstand harsh conditions. Al-Moameri et al. [5] utilized degradation models to predict the longevity of RPUFs, emphasizing their utility in applications that demand high environmental resilience [6].

Predictive modeling—encompassing reaction kinetics, thermomechanical, and multiscale techniques—has profoundly influenced RPUF research. These models drive enhancements in mechanical strength, thermal insulation, and fire resistance, advancing the development of RPUFs that meet diverse industrial demands and regulatory standards.

3.2 Raw Materials

3.2.1 Overview of the Effects

Raw materials are central to the formation and functionality of RPUFs, influencing their structural integrity and performance characteristics. Accurately representing these components in computational models is essential for predicting foam behavior and optimizing formulations. Computational modeling enables more accurate simulations that reflect real-world performance. This section explores how variations in polyols, isocyanates, blowing agents, catalysts, and fillers can influence foam properties and how computational models can optimize and innovate RPUF formulations. Representing raw materials in computational modeling is crucial for effectively synthesizing and optimizing RPUFs [10].

Polyols. It often begins with representation of the chemical structure of polyols, the backbone of polyurethane foam, including the molecular weight and reactive groups. The chemical characteristics of the polyol, including its functionalities, heat capacities, and hydroxyl values, were found to have a significant impact on the kinetics of the PU process [11]. Ugarte et al. [12] emphasize the importance of accurately modeling the hydroxyl functionality of polyols to predict its reactivity to isocyanates and final properties of the RPUFs. Hydroxyl number, which quantifies the number of hydroxyl groups in a polyol, is also a key parameter in modeling reactivity. A higher hydroxyl number typically indicates a more significant potential for crosslinking and, consequently, improved mechanical properties of the foam. Miao et al. [13] reported that synthesized polyols with high hydroxyl numbers (e.g., 317.0 mg KOH/g) can produce rigid polyurethanes with enhanced performance characteristics. Computational models can use hydroxyl number data to predict the extent of crosslinking and the mechanical properties of the final foam. Characterizing the hydroxyl number has the advantage of increasing the efficiency of describing a polyol and may also increase the precision of performance extrapolation [14]. The distribution of hydroxyl groups within the polyol structure also affects reactivity. For example, Souza et al. discussed how hydroxyl groups in different positions, such as terminal and pendant, can influence the reactivity of polyols with isocyanates [15]. The distribution of these hydroxyl groups affects the overall reactivity and RPUF's performance which can be evaluated through simulation.

The type and distribution of hydroxyl groups significantly influence the reactivity of polyols in the synthesis of RPUFs. Primary hydroxyl groups are generally more reactive than secondary hydroxyl groups, which can lead to differences in the foams' crosslinking density and mechanical properties. Ghoreishi et al. [14] conducted a comprehensive study on the reaction modeling of urethane polyols, focusing on the effects of primary, secondary, and hindered-secondary hydroxyl groups on reaction kinetics and temperature profiles. Their findings indicate that the presence of primary hydroxyl groups enhances the reactivity of polyols, leading to improved foam properties. Also, from Ghoreishi et. al.'s kinetic model, it reveals that the petroleum-based polyols have no primary hydroxyl groups. Petroleum-based polyols' lack of primary hydroxyls lessens polymerization reactivity, which may limit the crosslink density and slow down the rate of reaction. The foam's stiffness, dimensional stability, and thermal insulation qualities may be impacted by its reduced crosslink density, which makes it less appropriate for uses requiring great strength and deformation resistance [16]. Reaction kinetics modeling is crucial for predicting how different hydroxyl functionalities will affect the formation of urethane linkages. Ismail et al. explored the urethane-forming reaction kinetics of model palm olein polyols, quantifying the impact of primary and secondary hydroxyls on the reactivity with isocyanates. Polyols with primary hydroxyl groups have three times higher reaction rate constants and faster gel time than polyols with secondary hydroxyl groups [17].

Discussing Ghoreishi et al.'s model, the polyurethane gel process was investigated as being affected by different types of polyol. Equation (3.1) is the primary reaction in a polyurethane gel formation process and Table 3.1 is the summary of reactions with the key reactants in RPUF synthesis. The model accounts for primary, secondary, and hindered-secondary hydroxyls; by fitting the model to experimental data, the authors extract kinetic parameters that reflect the influence of the distribution of hydroxyl groups on the kinetics shown in Table 3.2.

$$RNCO + R'CH_2OH \rightarrow RNHCOOCH_2R' \tag{3.1}$$

$$Isoccyanate + Alcohol \rightarrow Urethane \tag{3.2}$$

Alcohol Group Reactions. Reactions 1–6 focus on the reactivity of different alcohol groups on monomer B. Primary alcohols (B_P) react more readily with isocyanate, initiating early urethane linkages, whereas secondary (B_S) and hindered-secondary (B_{HS}) alcohols react slower due to steric effects, which in turn modulates the foam's crosslinking density.

Polymer Urethane Linkages. Reactions 7–12 represent isocyanate groups reacting with primary, secondary, and hindered-secondary alcohol groups on the polymer chain (P_{BP}, P_{BS}, P_{BHS}). This crosslinking stage enhances the foam's mechanical properties and stability, affecting its structural integrity.

Formation of Allophanate or Biuret Linkages. Reactions 13 and 14 involve isocyanate and urethane groups, forming additional allophanate or biuret linkages, which contribute to RPUF's rigidity and temperature stability.

Water Reactions and Gas Generation. Reactions 15 and 16 involve water reacting with isocyanate to generate CO_2, which creates a cellular structure within the foam. This process determines the foam density and the overall cell structure.

Table 3.1 Summary of reactions

No	Reaction
1	$A + B_P \rightarrow P$
2	$A + B_S \rightarrow P$
3	$A + B_{HS} \rightarrow P$
4	$A + P_{BP} \rightarrow P$
5	$A + P_{BS} \rightarrow P$
6	$A + P_{BHS} \rightarrow P$
7	$B_P + P_A \rightarrow P$
8	$B_S + P_A \rightarrow P$
9	$B_{HS} + P_A \rightarrow P$
10	$P_A + P_{BP} \rightarrow P$
11	$P_A + P_{BS} \rightarrow P$
12	$P_A + P_{BHS} \rightarrow P$
13	$A + Ur \rightarrow P$
14	$P_A + Ur \rightarrow P$
15	$A + W \rightarrow N + CO_2$
16	$P_A + W \rightarrow P + CO_2$
17	$N + A \rightarrow P$
18	$N + P_A \rightarrow P$

A is isocyanate monomer, B is polyol monomer, P is polymer, B_P is primary alcohol group on monomer B, B_S is secondary alcohol group on monomer B, B_{HS} is hindered-secondary alcohol group on monomer B, P_A is isocyanate group on polymer, P_{BP} is primary alcohol group on polymer, P_{BS} is secondary alcohol group on polymer, PBHS is hindered-secondary alcohol group on polymer, N is amine, W is water, and Ur is urethane moiety

Amine-Catalyzed Reactions. In reactions 17 and 18, amine (N) catalysts speed up the reactions between isocyanates and hydroxyl groups, controlling cure rates and foam rise times, essential for optimizing the final properties of RPUF.

These reactions offer a detailed model that captures the nuanced effects of molecular structure, concentration, and reaction conditions on the kinetics of isocyanate-alcohol polymerization. These allow for precise control and prediction of the overall reaction process. Reaction rates for reference systems were used to obtain contour plots of gel temperature profiles for various polyols using computer simulation. These contour plots and kinetic parameters (Table 3.3) can provide information on the presence of primary, secondary, and hindered-secondary hydroxyl groups in a polyol. This simulation method offers helpful information on the polyol properties where current polyols can be used more effectively, creating new and enhanced polyols through molecular engineering [11].

Isocyanates. Polyurethane foams are built upon urethane linkages formed by a reaction between isocyanates and polyols. The chemical structure of isocyanates,

Table 3.2 Reaction rate expressions for the reactions presented in Table 3.1

Rxn. No.	k_0	E	h	Reaction rate expression
1	$k_0 (1)$	E (1)	h (1)	$r(1) = k(1) f_A C_A X_P f_B C_B$
2	$k_0 (2)$	E (2)	h (2)	$r(2) = k(2) f_A C_A X_S f_B C_B$
3	$k_0 (3)$	E (3)	h (3)	$r(3) = k(3) f_A C_A X_{HS} f_B C_B$
4	$k(10) = c_1 k(1)$		h (1)	$r(10) = k(10) f_A C_A (M_{BP} - X_P f_B C_B)$
5	$k(11) = c_1 k(2)$		h (2)	$r(11) = k(11) f_A C_A (M_{BS} - X_S f_B C_B)$
6	$k(12) = c_1 k(3)$		h (3)	$r(12) = k(12) f_A C_A (M_{BHS} - X_{HS} f_B C_B)$
7	$k(13) = c_1 k(1)$		h (1)	$r(13) = k(13) X_P f_B C_B (M_A - f_A C_A)$
8	$k(14) = c_1 k(2)$		h (2)	$r(14) = k(14) X_S f_B C_B (M_A - f_A C_A)$
9	$k(15) = c_1 k(3)$		h (3)	$r(15) = k(15) X_{HS} f_B C_B (M_A - f_A C_A)$
10	$k(22) = c_2 k(1)$		h (1)	$r(22) = k(22) (M_A - f_A C_A)(M_{BP} - X_P f_B C_B)$
11	$k(23) = c_2 k(2)$		h (2)	$r(23) = k(23) (M_A - f_A C_A)(M_{BS} - X_S f_B C_B)$
12	$k(24) = c_2 k(3)$		h (3)	$r(24) = k(24) (M_A - f_A C_A)(M_{BHS} - X_{HS} f_B C_B)$
13	$k_0 (25)$	E (25)	h (25)	$r(25) = k(25) f_A C_A C_{Ur}$
14	$k(26) = c_3 k(25)$		h (25)	$r(26) = k(26) C_{Ur}(M_A - f_A C_A)$
15	$k_0 (27)$	E (27)	h (27)	$r(27) = k(27) f_A C_A C_W$
16	$k(28) = c_3 k(27)$		h (27)	$r(28) = k(28) C_W(M_A - f_A C_A)$
17	$k_0 (29)$	E (29)	h (29)	$r(29) = k(29) f_A C_A C_U$
18	$k(29) = c_4 k(29)$		h (29)	$r(30) = k(30) C_U (M_A - f_A C_A)$

k_0 and E are Arrhenius equation constants and h is the enthalpy of reaction. Subscripts are P (primary), S (secondary), HS (hindered-secondary), B (alcohol moiety), and A (isocyanate moiety). The term c_i indicates constants

Table 3.3 Summary of kinetic parameters of primary, secondary, and hindered-secondary hydroxyls

Parameter	k_0	E	h	U
Primary	500	37,000	68,000	2
Secondary	55	40,000	68,000	2
Hindered-secondary	42	40,000	68,000	2

including the types and arrangements of functional groups, strongly affects their reactivity in this process. Their reactivity is influenced by their chemical structure, including the presence of functional groups. Zieglowski et al. investigated the influence of different isocyanate structures on the properties of polyurethane foams made from lignin-based polyols. They found that the reactivity of isocyanates, such as diphenylmethane-diisocyanate (MDI) and toluene diisocyanate (TDI), significantly affected the mechanical properties of the resulting foams [18]. The kinetics of the reaction of isocyanate with polyols and water is significant in understanding the process of foam expansion. Grzęda et al. discussed how isocyanates react with water, causing it to produce carbon dioxide, which further causes foaming [19]. The modeled reaction kinetics is one of the most crucial steps for the prediction of the rate of foam expansion and the final density of the foam. Incorporating kinetic parameters with computational models will help the researchers optimize formulations toward the desired performance characteristics. The isocyanate index also influences the kinetics of the reaction of isocyanates and polyols. Zou et al. highlighted that an increase in isocyanate content can accelerate the crosslinking reaction, thereby reducing the foam's tack-free time [20]. This information is essential for computational models that aim to predict the timing and efficiency of the foaming process and its final properties.

One kinetic modeling technique [21] described the reactions of isocyanates, such as MDI and TDI, with selected alcohols that mimic the chain ends of polymers used in PUFs. Using high molar ratios of alcohols to diisocyanates, MDI reacts about 1.5 times faster than TDI, while TDI shows a more complicated kinetic situation. This high reactivity and ability to create a dense and well-crosslinked network benefit RPUF synthesis, enhancing structural integrity and thermal insulation. Using a titration approach, Ni has conducted kinetic research using linear regression to estimate the rate constants of the isocyanate-water reaction and calculate the concentration of the isocyanate group as a function of time [22]. Li et al. added that a higher concentration of isocyanate within the polymer network of polyurethane is directly correlated with an improved crosslink density. This would mean that more significant amounts of isocyanate yield greater crosslinkage between the polymer chains, creating a more rigid network through densification. Increased crosslinking led to higher mechanical or compressive strengths with increased rigidity, contributing to the strength needed for structural stability in specific applications, such as rigid foams used as insulation [23]. These findings of accurately modeling the kinetics of isocyanates in computational frameworks can help predict the behavior of PUFs during the curing process.

Zhao and Suppes' study provides valuable insights into the effect of different isocyanates on reaction enthalpies and the formation of urethanes, specifically through computational modeling. Computational modeling indicates that the size and location of functional groups in isocyanate molecules significantly impact reaction enthalpies. MDI and TDI were compared to assess how molecular size affects enthalpy of reaction. Figure 3.1 shows the isocyanate structure used in their study.

By computational method using Gaussian 09 software (molecular modeling), kinetic and thermodynamic parameters utilized in simulation research were enhanced

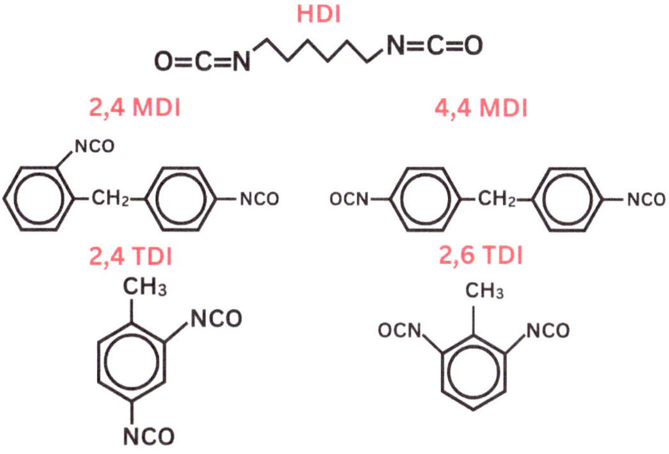
HDI
O=C=N
N=C=O
2,4 MDI
4,4 MDI
NCO
CH2
NCO
OCN
CH2
NCO
2,4 TDI
2,6 TDI
CH3
NCO
CH3
OCN
NCO
NCO

Fig. 3.1 Structures of various isocyanates

and the recommended values, presented in Table 3.4, were obtained. Larger isocyanate molecules, such as methylene diphenyl diisocyanate (MDI) versus toluene diisocyanate (TDI), exhibit lower reaction enthalpies, suggesting that increasing molecular size reduces energy release during the reaction. Additionally, steric hindrance affects reactivity. For example, 2,4-TDI versus 2,6-TDI has different reaction enthalpies due to varying steric interactions at different functional group positions [24].

Blowing agents. The role of blowing agents in the computational modeling of RPUF formation is crucial for understanding how these agents influence the physical properties and performance of the final product. Blowing agents generate the gas that expands the foam, creating its cellular structure. Most foaming material comes from the blowing agent used in the foaming process, the liquid phase, or the reaction or breakdown of chemical material under the influence of heat or a catalyst. The classifications for blowing agents include chemical or physical blowing agents based on how they are released during the foaming process [25]. The physical or chemical blowing agent's evaporation is controlled by the vapor–liquid partition of the volatile compound between the vapor and the polymeric phases, which can also affect the

Table 3.4 Recommended values for heat of reaction (kJ/mol)

	PMDI	MDI	TDI	HDI
Primary	−82.0	−84.5	−87.2	−86.7
Secondary	−78.8	−81.2	−84.0	−83.7
Tertiary (HS)	−68.7	−70.8	−74.2	−73.5
Water	−81.1	−83.6	−85.8	−67.3
Amine				−67.4

entire foam-growing process [26]. The choice and concentration of blowing agents mainly influence the density, mechanical strength, and thermal insulation properties of RPUFs. Jaf et al. discussed the change in the type and amount of blowing agent used, which influences the optimization of polyurethane foam performance concerning variations in foam density and structural integrity [27]. Computational models can simulate these relationships, allowing researchers to predict how changes in blowing agent formulation will impact the final foam characteristics.

The behavior of blowing agents during the foaming process can be complex, involving mass transfer, vaporization, and gas diffusion. Al-Moameri et al. developed a computer-based simulation to model the performance of various blowing agents, including their evaporation rates and effects on cell morphology. The authors propose a model explaining evaporation as a function of an overall mass transfer coefficient, which is affected by the variation in the vapor phase activity of the blowing agent inside the foam cells as opposed to the cell walls made of resin. This method enables a more accurate depiction of the foaming dynamics, considering various factors inside the foam structure. The study found that successful modeling of the foaming process hinges upon using a mass transfer coefficient that decreases to near zero as the foam resin approaches its gel point [28]. Al-Moameri et al. compared their simulation model with experimental data for the six blowing agents under study to validate it. The efficiency of the modeling technique in capturing the dynamics of the foaming process is demonstrated by the successful correlation between the experimental findings and the model predictions. Confidence in the model's prediction skills, which can be used to optimize formulations for particular applications, depends on this validation [28].

Computational models, such as those developed in the study by Al-Moameri et al. [29], simulate both physical (e.g., n-pentane) and chemical blowing agents (e.g., water) within the polyurethane foam matrix. Through simulations, these models can predict how the blowing agents will behave in different formulations by tracking their mass transfer and diffusion rates into foam cells.

In order to better understand blowing agent mass balance in rigid foams, this paper uses a MatLab simulation algorithm for the urethane foaming process. As Al-Moameri et al. [28] previously explained, the simulation code is based on fitting parameters of kinetic and physical properties. The simulation code's algorithm is displayed in Fig. 3.2. The ODE45 function solves the differential equations of the rate of change in reaction temperature, concentrations, foam height, the rate of evaporation of the physical blowing agent, and the mass transfer rates of blowing agents from the resin to the cells after the recipe, initial conditions, and database parameters are entered into the FoamSim code. Simulations indicated that water (a chemical blowing agent) is highly effective (>85%) due to rapid CO_2 generation. In contrast, physical agents like n-pentane show lower effectiveness, with part of the agent often getting trapped in the resin [29].

Jaf et al. created MATLAB code to model how the performance of RPUF might be affected by physical and chemical blowing agents at various loadings. This kind of computational modeling lowers the time and expense of experimental testing

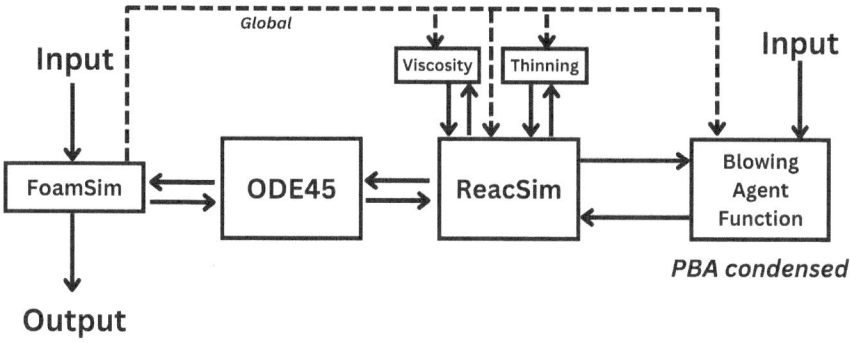

Fig. 3.2 Algorithm of the simulation code

while improving the accuracy of outcomes. Researchers can more effectively optimize foam formulations by simulating the foaming process and forecasting how formulation modifications will impact the final foam properties [27]. Water was used as a chemical blowing agent, and physical blowing agents such as n-pentane, cyclohexane, and methyl formate were also assessed. The unique qualities of each of these substances affect the foaming process and the final foam's attributes. For example, while foaming, physical blowing agents usually evaporate, producing gas bubbles that cause the foam to expand. On the other hand, foam expansion is facilitated by carbon dioxide production through the reaction of chemical blowing agents with isocyanates [27]. The results indicated that different blowing agents lead to variations in cell structure and size, which are essential for determining the foam's overall performance.

Catalysts. The role of catalysts in the formation of RPUF is crucial for optimizing the synthesis process and enhancing the foam's properties. Catalysts facilitate the reaction between isocyanates and polyols, influencing the foam's mechanical, thermal, and chemical properties. The two main categories of catalysts utilized in producing polyurethane foams are amine and tin-based catalysts. Amine catalysts, such as tertiary amines, are commonly used to accelerate the reaction between isocyanates and polyols. Mahmoud et al. noted that amine catalysts help precisely control the relative reaction rates of isocyanate with both polyol and water, which is crucial for achieving the desired foam structure and properties [30]. Tin-based catalysts, such as dibutyltin dilaurate (DBTDL), facilitate a good reaction between isocyanates and polyols with a faster curing time. However, using these catalysts has resulted in environmental concerns due to toxicity, which has led to the search for alternative catalysts [31]. The catalyst's presence affects the reaction's kinetics during the foaming process. According to Wang et al., isocyanate and polyol reaction rate imbalance can lead to foam collapse or inappropriate cell structure formation [32]. Catalysts regulate the rates of two critical reactions in polyurethane foam synthesis: gelling (forming urethane linkages) and blowing (generating gas from the reaction with water). This regulation is essential for achieving the desired foam characteristics, including density and mechanical strength [30].

Catalysts are critical in balancing the gelling (isocyanate-polyol) and blowing (isocyanate-water) reactions. Computational models that precisely incorporate the catalytic rates of these reactions enable better control over foam expansion, cell structure, and density. For example, the study discovered that by investigating the gelling and blowing kinetic rates of RPUFs, compared to a single-catalyst system, a combination catalyst system (DMCHA + DABCO-33LV) with a 1:1 mass ratio and 0.35% w/w loading increased the foam's compressive strength by 26.4% [33]. A MATLAB model was created by Zhao et al. [34] based on the simultaneous solution of differential equations to incorporate the effect of catalyst loading on polyurethane reactions and physical processes. The model parameters for this thermoset polymerization were fitted through experiments with varying catalyst loading levels. Rate constants for the foaming process of rigid polyurethane were based on reaction temperatures and catalyst types. Equations 3.3 and 3.4 show the model equation that accurately accounts and simulates the effects of various catalysts.

$$r_{\text{gel}} = (k + k_{\text{cat8}} \cdot c_{\text{cat8}} + k_{\text{cat5}} \cdot c_{\text{cat5}}) \cdot c_{\text{iso}} \cdot c_{\text{OH}} \tag{3.3}$$

$$r_{\text{blow}} = (k' + k'_{\text{cat8}} \cdot c_{\text{cat8}} + k'_{\text{cat5}} \cdot c_{\text{cat5}}) \cdot c_{\text{iso}} \cdot c_{\text{OH}} \tag{3.4}$$

where

- k and k' represent the gel and blow reaction rate;
- k_{cat8} and k'_{cat8} represent the gel and blow reaction rate with catalyst 8 (DMCHA);
- k_{cat5} and k'_{cat5} represent the gel and blow reaction rate with catalyst 5 (PMDETA).

Separate gel reactions using three different types of polyol were employed to collect temperature data over time. There are four components to single polyol gel reactions. In the first, the reaction is carried out without catalysts; in the second, only catalyst 5 is present; in the third, only catalyst 8 is present; and in the final, only catalyst 8 is present, but in different amounts. Based on experimental results, all single polyol kinetics parameters were fitted. Table 3.5 lists all the specific values. Both Cat8 and Cat5 greatly influence the internal temperature of the gel reaction. An increasing amount of catalyst accelerates the gelling reaction, resulting in a higher and earlier temperature rise [34].

Blowing agents (water and methyl formate) and blowing catalysts were introduced to the gel formation to produce rigid foams and track the foaming process. In contrast to those published by Tesser et al. [26], Table 3.6 presents the blow reaction kinetic parameters concerning various catalytic conditions. As can be seen, significantly identical values for the pre-exponential factor and activation energy were discovered, suggesting that the kinetics behaved very similarly.

This model can forecast the performance of foams with various combinations of isocyanate, polyols, catalysts, chemical blowing agents (like water), and physical blowing agents (like methyl formate) in recipes where behavior is based on

Table 3.5 Kinetic parameters of the gel reaction in relation to various catalytic conditions

Poly G76-635	Without catalyst	Catalyst 8	Catalyst 5
k (ml/(mol*s*g catalyst)	0.7125	51	60
E (J/mol)	50,000	50,000	43,000
H (J/mol)	79,000	79,000	79,000
Poly V360	*Without catalyst*	*Catalyst 8*	*Catalyst 5*
k (ml/(mol*s*g catalyst)	0.7	50	42
E (J/mol)	48,000	41,000	34,000
H (J/mol)	73,000	73,000	73,000
Poly R315x	*Without catalyst*	*Catalyst 8*	*Catalyst 5*
k (ml/(mol*s*g catalyst)	200	300	400
E (J/mol)	10,000	10,000	10,000
H (J/mol)	67,000	67,000	67,000

Table 3.6 Kinetic parameters of blow reactions with respect to various catalytic conditions

Poly G76-635	Without catalyst	Catalyst 8	Catalyst 5	Reference
Iso-water				
k (ml/(mol*s*g catalyst)	60	200	200	25.8
E (J/mol)	50,000	50,000	50,000	44,100
H (J/mol)	67,000	67,000	67,000	86,000
Iso-amine				
k (ml/(mol*s*g catalyst)	150	150	150	–
E (J/mol)	40,000	40,000	40,000	–
H (J/mol)	20,000	20,000	20,000	–

pure component kinetic parameters (e.g., single polyol performance). The simulation model can speed up the creation of new foam formulations, mainly when novel bio-based polyols are included in formulations.

Surfactants. Polyurethane foaming systems use silicone surfactants composed of random copolymer grafts of polyethylene oxide-co-propylene oxide (PEO-PPO) and a polydimethylsiloxane (PDMS) backbone. Cell size control to produce a fine-celled structure with a limited cell size distribution and emulsification are the main functions of silicone surfactants in rigid foam formulations [35]. In the study of Zhao et al. [36], to create homogeneous foam cells and avoid coalescence, they describe how surfactants lessen surface tension at the air-resin interface, promoting bubble nucleation, growth, and stabilization. Surface tension and bubble stability parameters are used to model this stabilization effect, which aids in simulating realistic cell structures within the foam. The paper uses several equations to model these effects. For example, they calculate the number of nucleation sites, N_c, and bubble radius, r, based on surface tension, σ, bubble pressure, and energy input W:

$$N_c = \frac{W}{4\pi r^2 \sigma} \tag{3.5}$$

By altering the mixing time and concentration of surfactants in the model, the simulation could predict how these factors would affect the size of bubbles and the uniformity of foams. Fewer nucleation sites were formed with smaller surfactant concentrations and shorter mixing times, producing more significant, irregularly shaped bubbles. Fine cell structure was attained when larger amounts of surfactant with longer mixing times were applied, meaning that careful balancing was critical to achieving desired foam characteristics. By incorporating the surfactant's surface tension effect, the simulation predicted that higher surfactant concentrations lower the liquid's surface tension. This reduction stabilizes the bubbles, preventing them from merging prematurely (coalescence) and allowing them to grow uniformly. The model demonstrated that lower surface tension correlates with increased foam stability, as bubbles remain intact longer during polymerization [36]. The algorithm used to determine the bubble radius during the foaming process is displayed in Fig. 3.3.

Fillers. Computational models can simulate the effects of fillers, such as their size, distribution, concentration, and interaction with the PU matrix, and help predict foam properties like density, compressive strength, thermal insulation, and water resistance. The study by Sharma and Prasath Kumar [37] leverages machine learning, specifically Linear Regression (LR) and Artificial Neural Networks (ANN), to predict the physico-mechanical properties of sandwich panels made from PUF infused with

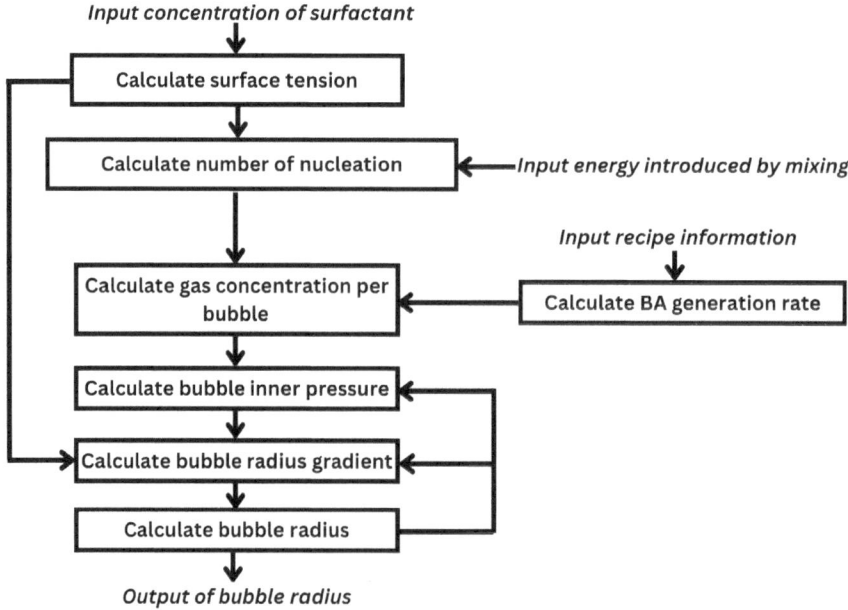

Fig. 3.3 Bubble radius calculation algorithm for the foaming process

Fig. 3.4 ANN flowchart for predicting sandwich panel compressive and flexural strength

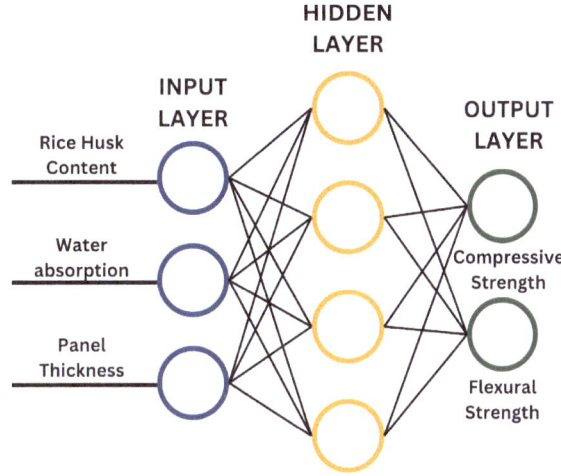

rice husk (RH) as a filler shown in Fig. 3.4. By varying RH content (10–50%) and core thicknesses (2 cm and 3 cm), the researchers constructed panels with calcium silicate board (CSB) face sheets and assessed their compressive and flexural strengths through experimentation. They trained the LR and ANN models on these mechanical properties using a full factorial design for comprehensive data collection. They validated them with the coefficient of determination (R^2) and prediction error metrics. The ANN model demonstrated superior accuracy, achieving R^2 values above 0.90 and errors below 4%, outperforming LR. Findings indicated that 30% RH content optimized the panels' mechanical properties, demonstrating the utility of machine learning in designing sustainable construction materials with natural fillers like rice husk [37].

3.2.2 Role of Hydroxyl Moieties

The polymerization of rigid polyurethane foams (RPUFs) involves the reaction between polyols, which contain hydroxyl (OH) groups, and isocyanates. These hydroxyl groups play a crucial role in determining the foam's structure and mechanical properties. The reactivity and concentration of hydroxyl moieties within polyols directly influence the degree of crosslinking in the resulting polymer network. Crosslinking refers to the formation of bonds between polymer chains, which is key to the foam's strength, stability, and performance. The hydroxyl groups facilitate these reactions by reacting with isocyanate groups to form urethane linkages, a process that ultimately governs the polymer's molecular structure and, consequently, its macroscopic properties.

Reactivity and Concentration of Hydroxyl Groups. The hydroxyl groups present in polyols can be classified as primary, secondary, or hindered-secondary, based on their chemical structure. Each type of hydroxyl group exhibits different reactivity when interacting with isocyanates, leading to varying levels of crosslinking and different mechanical properties in the final foam product.

1. *Primary Hydroxyl Groups.* These groups are the most reactive because their hydroxyl hydrogen atom is readily available to react with isocyanates. The presence of primary hydroxyl groups accelerates the polymerization process, leading to the rapid formation of urethane linkages and a higher crosslinking density. A higher crosslinking density generally results in a more rigid and stronger foam. Polyols with a high concentration of primary hydroxyl groups, therefore, produce RPUFs with improved mechanical properties such as higher tensile strength, compressive strength, and thermal stability [14].
2. *Secondary Hydroxyl Groups.* Secondary hydroxyl groups are less reactive than primary hydroxyls due to the steric hindrance caused by the adjacent carbon atoms. As a result, the polymerization process is slower when secondary hydroxyls are involved, leading to lower crosslinking density. This can produce foams that are more flexible and less dense. Although these foams may have some desirable properties in terms of flexibility, they typically exhibit lower mechanical strength compared to those made from polyols with a higher content of primary hydroxyl groups [8].
3. *Hindered-Secondary Hydroxyl Groups.* Hindered-secondary hydroxyls are even less reactive due to the significant steric hindrance in their structure, which makes them much slower to react with isocyanates. The reduced reactivity of hindered-secondary hydroxyl groups results in a much lower crosslinking density, which generally leads to foams with poor mechanical properties. These groups are often found in bio-based polyols that have undergone extensive functionalization to enhance their performance in various applications. However, the lower reactivity associated with hindered-secondary hydroxyls can be a disadvantage when aiming for high strength, rigid polyurethane foams [2].

Influence of Hydroxyl Moieties on Crosslinking Density and Mechanical Properties. The crosslinking density is a critical determinant of the mechanical properties of RPUFs. A higher concentration of reactive hydroxyl groups, particularly primary hydroxyls, leads to a more tightly crosslinked polymer network. This network provides greater rigidity and higher mechanical strength, as well as improved thermal stability. Conversely, a lower crosslinking density results in a more flexible foam, which may be desirable in some applications but not in those requiring high strength or durability, such as in thermal insulation or structural components [38]. For instance, Alfeche et al. [2] found that RPUFs made from polyols with a higher fraction of primary hydroxyl groups exhibited significantly higher compressive strength and thermal stability compared to those made from polyols with a higher fraction of secondary or hindered-secondary hydroxyl groups. Similarly, Ghoreishi et al. [14] demonstrated that the mechanical properties of polyurethane foams could be tailored by controlling the distribution of hydroxyl groups in the polyol component. Foams with a balanced mix of primary and secondary hydroxyl groups exhibited desirable mechanical properties, combining some degree of flexibility with increased strength.

In practical terms, a higher hydroxyl group concentration typically enhances the density and integrity of the foam, making it more suitable for applications such as insulation panels, automotive parts, and construction materials, where strength and durability are crucial. Conversely, foams with lower hydroxyl group concentrations may be used in applications where lower strength and flexibility are more important, such as in cushions or packaging materials.

Influence of Polyol Type on Hydroxyl Moieties and Foam Performance. Polyols for RPUF production can be derived from petrochemical sources or bio-based feedstocks, and the type of polyol used has a significant impact on the structure of the hydroxyl moieties and, consequently, the properties of the foam.

Petrochemical Polyols. Petrochemical-based polyols, such as those derived from propylene oxide (PO) and ethylene oxide (EO), typically contain a mixture of primary and secondary hydroxyl groups. These polyols are well-established in the polyurethane industry due to their consistency and predictable reactivity, which allows for the reliable production of high-performance foams. The concentration of primary hydroxyl groups in petrochemical polyols can be adjusted through the manufacturing process, enabling precise control over the polymerization rate and crosslinking density (Zhao et al. [8]). This flexibility is advantageous when designing foams with specific mechanical properties for various applications, from rigid insulation to automotive parts. Polyurethane foams made from petrochemical polyols generally exhibit excellent mechanical properties, such as high tensile strength and compressive strength, making them ideal for structural applications. Additionally, the high reactivity of primary hydroxyl groups in these polyols ensures fast gelation times, which is beneficial in industrial foam production processes [14].

Bio-Based Polyols. Bio-based polyols, derived from renewable resources such as vegetable oils, lignocellulosic biomass, or even algae, often contain a higher proportion of secondary or hindered-secondary hydroxyl groups. These polyols are considered more sustainable alternatives to petrochemical polyols but tend to exhibit slower reaction rates due to the reduced reactivity of their hydroxyl groups. For instance, coconut oil-based polyols and soy-based polyols are examples of bio-based polyols that have been extensively studied for use in RPUF formulations [2, 38]. The increased presence of secondary and hindered-secondary hydroxyl groups in bio-based polyols can lead to foams with lower crosslinking density and, therefore, reduced mechanical strength compared to their petrochemical counterparts. However, bio-based polyols offer significant advantages in terms of sustainability, with many of them being biodegradable and derived from renewable resources. Additionally, the presence of secondary hydroxyl groups in bio-based polyols can introduce beneficial properties, such as increased flexibility or improved resistance to certain environmental factors, making them suitable for specialized applications [38].

3.2.3   Case Studies and Experimental Data

Influence of Hydroxyl Group Distribution on Foam Properties. A study by Mendija et al. [38] investigated the effect of polyol hydroxyl group distribution on the mechanical properties of RPUFs. Polyols with different ratios of primary to secondary hydroxyl groups were synthesized and used to produce foams. The results indicated that foams made from polyols with a higher fraction of primary hydroxyl groups exhibited superior mechanical properties. Specifically, foams with a primary hydroxyl content of 70% showed a compressive strength of 0.5 MPa, while those with 30% content had a compressive strength of 0.3 MPa. This demonstrates that increasing the primary hydroxyl content enhances the foam's mechanical strength [38].

Effect of Polyol Type on Foam Performance. Alfeche et al. [2] compared RPUFs produced from petrochemical-based polyols and bio-based polyols derived from coconut oil. The study found that foams made from bio-based polyols with a higher fraction of primary hydroxyl groups exhibited improved mechanical properties. For instance, a foam produced from a bio-based polyol with 60% primary hydroxyl content achieved a tensile strength of 1.2 MPa, whereas a foam from a polyol with 40% primary hydroxyl content had a tensile strength of 0.8 MPa. This highlights the importance of hydroxyl group distribution in determining foam performance, even when using renewable resources [2].

Computational Modeling of Hydroxyl Group Effects. Ghoreishi et al. [14] developed a computational model to predict the impact of hydroxyl group distribution on the reaction kinetics and mechanical properties of polyurethane foams. The model incorporated parameters such as the fraction of primary, secondary, and hindered-secondary hydroxyl groups. Simulations indicated that increasing the fraction of primary hydroxyl groups from 50 to 70% resulted in a 20% increase in foam density and a 15% improvement in compressive strength. This underscores the predictive capability of computational models in understanding and optimizing foam properties based on polyol chemistry [14].

Impact on Computational Modeling of RPUFs. The influence of hydroxyl moieties on the polymerization process of RPUFs can be effectively modeled using computational tools. By incorporating the concentration and type of hydroxyl groups into the reaction kinetics models, it is possible to predict the properties of the resulting foam without the need for extensive experimental trials. Computational models for RPUFs typically involve solving a set of differential equations that describe the reaction between isocyanates and hydroxyl groups, the dynamics of foam expansion, and the heat generation during the foaming process. The concentration of primary, secondary, and hindered-secondary hydroxyl groups can be integrated into these models to simulate the effect of different polyol formulations on foam performance. This allows for a more efficient design process, where the properties of the foam can be predicted and optimized before physical production begins [8, 38].

The hydroxyl moieties in polyols play a central role in determining the properties of rigid polyurethane foams. The concentration and type of hydroxyl groups, whether primary, secondary, or hindered-secondary, directly affect the crosslinking density of the polymer network, which in turn influences the foam's mechanical strength, thermal stability, and overall performance. While petrochemical polyols typically offer faster reaction rates and higher mechanical strength, bio-based polyols provide a more sustainable alternative, albeit with some trade-offs in terms of foam rigidity and strength. By carefully selecting the type of polyol and optimizing the hydroxyl group distribution, it is possible to design foams with tailored properties for a wide range of applications. Computational modeling provides a powerful tool for predicting foam performance based on the distribution of hydroxyl groups in the polyol. This approach allows for more efficient foam formulation and optimization, reducing the need for extensive experimental work and contributing to the development of more sustainable and high-performance polyurethane materials.

3.3   Thermo-Kinetics

3.3.1   Correlation Between Thermo-Kinetic Parameters and Mechanical Properties

Rigid polyurethane foams (RPUFs) are a class of versatile materials with a wide range of applications [39–41]. They are used in construction, appliances, and transportation as a robust insulating material. In the pursuit of developing sustainable and more efficient customer-focused RPUFs or manufacturing methods thereof, turning to the underlying physical processes involved in RPUF formation is the wisest first course of action. In this regard, computational techniques offer significant utility.

Computational simulations based on reaction kinetics and heat transfer differential equations have been performed on polyurethane (PU)-forming processes [41–58]. For example, Krol et al. developed a software package for linear PU systems that can generate 2D diagrams of concentration and conversion changes [41]. Zhao and colleagues created a simulation method to estimate dynamic concentrations, degree of polymerization, blowing agent evaporation, and other relevant parameters [42–44]. Rusche et al. introduced MoDeNa, a multiscale simulation framework that unifies several existing software packages and provides outputs such as bubble growth rate, foam density, conversion of components, reaction temperature profile, and bubble radius [45]. Raimbault et al. developed FOAMAT®, a device for measuring physical parameters like dynamic foam height, reaction temperature, and gas content in PU foam systems [46]. Wright et al. created FoamPi, an open-source Raspberry Pi device for monitoring polyurethane foaming reactions, measuring temperature rise, foam height changes, and mass variations [47]. While these tools offer valuable insights, they lack a strong focus on the final PU foam properties that are crucial for the PU foam's final application. Additionally, many of these models and their validations are based on commercial polyols, limiting their applicability to new polyols.

The overall RPUF formation is a complex interplay of various reactions and their interactions (see Fig. 3.5) [39–41]. For example, a blowing reaction can be made to happen simultaneously with the PU formation/gelling or the isocyanate-polyol reaction. In water-blown RPUF formation, the reaction between an available isocyanate and water produces gaseous carbon dioxide (CO_2) responsible for the foam (volume) expansion. Other reactions that can also take place are the reaction between an available isocyanate and amine moiety produced from the water-blown RPUF formation or from the polyol compound used for the RPUF formation, and produce a urea compound. With such complexity, precise control over the reaction is crucial to achieve the desired properties and behavior of the final RPUF product.

The onset of stable network formation within a polyurethane system, primarily driven by the rapid formation of urethane linkages, is marked by the gel time of the PU-forming process [39–41]. It is a parameter reflective of the complex thermo-kinetics of the various reactions (i.e., their heats involved, their %conversion over time, etc.) during the PU formation. The parameter is influenced by factors like the nature of reactants used, their compositions in the reacting mixture, their mixing process, the initial temperature at which they were mixed, the surrounding temper-ature, etc. In this regard, the variability of conditions introduces variability to the

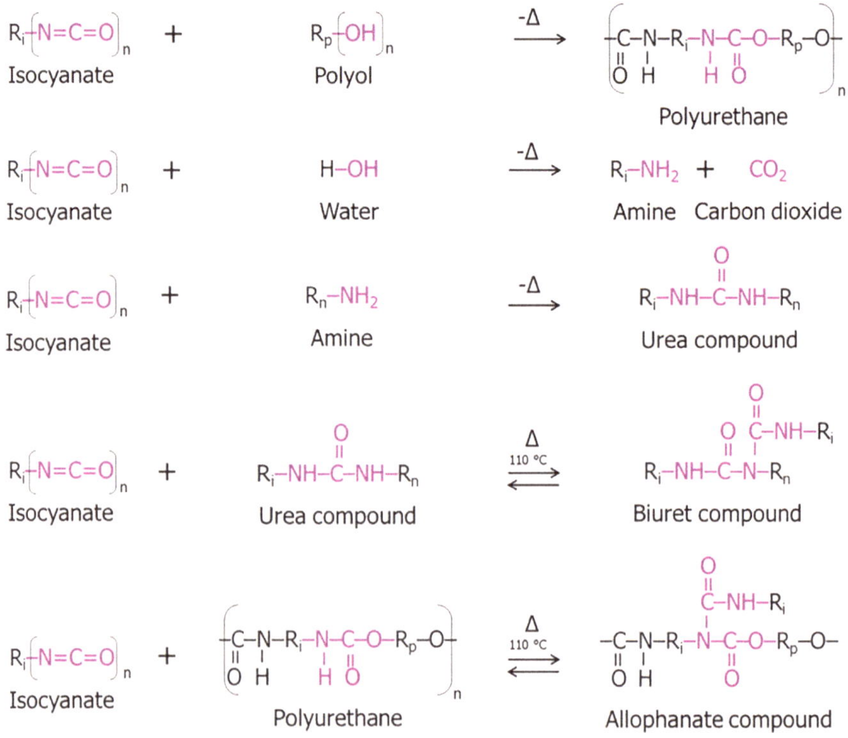

Fig. 3.5 Reactions that can occur in polyurethane foam synthesis

parameter as well. Moreover, it is important to remember that, for the formation of cellular foams, while urethane linkages are forming, concurrent evolution of gases, facilitated by blowing agents, is essential. However, if the blowing reaction proceeds too rapidly, the formation of a stable network may be compromised due to the premature release of gases, preventing the robust or effective crosslinking and solidification of the cellular polymer structure, thus inferior mechanical properties such as compressive strength [34, 48, 49]. Conversely, if the blowing reaction proceeds too slowly, the gas evolution rate may not be able to keep pace with the formation of solid network/s, resulting in a denser and less voluminous material, with lesser thermal and sound-insulating prowess [2, 13, 28, 29]. With the complex interplay of the blowing and gelling reaction, authors have hypothesized that there must exist an optimal gel time at which the rate of network solidification and the rate of gas evolution would ensure the development of a structurally sound cellular PU foam [2].

Using the gel time of PU formation, Alfeche et al. pioneered designing high-performance RPUFs from new polyols via computational technique using the parameter [2]. They were able to do so by using simulated temperature profiles reflective of the complex thermo-kinetics during the PU formation. They found that, for new polyols, a PU formulation that will yield a gel time closer to the gel time of a formulation using a conventional polyol results in a cellular foam with a better compressive strength and less shrinkage phenomena. More particularly, in their study, direct substitution of the polyol in a conventional formulation, which led to a gel time of 195 s, did not yield desirable results. The formulation exhibited a compressive strength of only 0.3 MPa and a significant shrinkage of 67% after three days compared to their control whose compressive strength was at 2.0 MPa and whose shrinkage after three days was at 4%. In contrast, a specific formulation using the new polyol, which resulted in a gel time of 315 s (within 5% of the control's gel time), demonstrated substantially improved performance. The formulation achieved a compressive strength of 1.1 MPa and a shrinkage of only 5% after three days.

Furthermore, Alfeche et al. were able to alter the gel time of the PU formation of their unique polyol by playing with the rate of catalyzed gel formation [2]. Catalysts are used in PU formation to ensure that less energy and cost is utilized for effective mixing as the isocyanate-polyol reaction requires a considerable activation energy of 39,000–54,000 J/mol [5, 43–45]. To overcome the limitations of their current RPUF using their unique polyol, the authors pursued to find, by computational means, the optimal condition for the catalyst loading in the PU-forming mix that comprises their unique polyol. With their unique polyol exhibiting an intrinsic autocatalytic activity, the authors assumed a 2% increase of the kinetic parameters of the urethane-forming reactions in their computational models. The models were then used to simulate temperature profiles and find the formulation with a catalyst loading that will yield a gel time within 5% of their control's, considering that such constraint on the gel time would ensure a balanced gelling and blowing reaction. The simulation of temperature profiles was done iteratively for theoretical mixes with catalyst loadings from 0.00 to 2.00 g simulated at 0.005 intervals until a valid and sound result was found. Thereafter, experimental validation of the computationally found formulation on the valid and sound catalyst loading in a PU formulation using the unique polyol showcased improvement of the resulting foam's compressive strength and shrinkage phenomena and elasticity as earlier discussed. More specifically, the authors achieved a compressive strength increase of >300% and three-day shrinkage decrease of >60% using their computationally derived formulation considering changes, assumptions, or constraints in OHV, hydroxyl moieties, kinetic parameters, or gel time, against a formulation that considers only a direct substitution of the raw materials used. Their findings provide further evidence of the role of precise reaction control in determining the final properties and performance of PU materials and present opportunities for further research and innovation in the field and groundbreaking discoveries.

3.4 Heuristics

3.4.1 Methods and Assumptions Employed

Heuristics are simplified approaches used to solve complex problems. They reduce the computational cost of simulations (run-time, coding & debugging time, etc.), and are especially helpful in solving problems with high degrees of freedom or highly constrained problems with limited empirical data or information.

In PU modeling common assumptions and heuristics include those outlined in Table 3.7.

While these assumptions offer time savings, they introduce limitations to the accuracy of predictions made and models used. For example, the assumption that the

Table 3.7 Some common heuristics used to simplify the computation of the temperature profile of a polyurethane foam formation

No	Assumption/Heuristic	References
1	The reactivity and heat of reaction of unreacted functional groups remain unchanged when they become part of a polyurethane polymer	[2, 38, 42–44, 51]
2	Hydroxyl groups (primary, secondary, hindered-secondary/tertiary) have different reactivity but the same heat of reaction	[2, 38, 42–44, 51]
3	Surfactants have minimal impact on reaction kinetics and thermodynamics	[2, 11, 38, 51]
4	Surfactants and catalysts do not interact	[2, 38, 42–44, 51]
5	Catalysts have minimal impact on reaction thermodynamics	[2, 38, 42–44, 51]
6	Concentrations are based on the initial total volume of liquid/resin, assuming an ideal mixture	[2, 38, 42–44, 51]
7	The heat capacity of the mixture is calculated as the ideal sum of component heat capacities, which are linear over the temperature range of RPUF formation	[2, 38, 42–44, 51]
8	All reactions exhibit 1:1:1 Isocyanate/Polyol/Polyurethane equivalents stoichiometry	[2, 38, 42–44, 51]

reactivity of unreacted functional groups remains constant upon incorporation into a polymer chain is overly simplistic. The reactivity of a reactant or the rate at which their reaction with other chemical species occurs is controlled by factors like their concentration in the mix, physical state including the spatial freedom of reacting groups, the temperature of the reaction, etc. which are all dynamic. While authors have considered accounting for the effect of dynamic concentration and temperature changes in a reacting mix in their simulations, the dynamic changes in the spatial freedom of reacting groups as the reaction proceeds have received limited attention. The simplification can be a source of error because as the polymer grows, it develops multilevel structures that provide steric effects that can influence the number of effective collisions for a reaction among chemical moieties to occur. Another over-simplification common in PU modeling is neglecting surfactants-catalyst interaction. Surfactants play a crucial role in PU synthesis by influencing the formation and stabilization of the polymer microstructure. Neglecting the interaction between surfactants and catalysts can be a source of error or inaccuracies in PU modeling for several possible reasons. These include possible surfactant-induced mechanisms, such as synergy, where surfactants may also impart catalytic activity and/or vice versa, or where surfactants may cause catalysts to aggregate or be entrapped in or partitioned into micelles and separate phases that can lead to uneven distribution within the reacting mixture or localized catalysis, resulting in complex variations in the polymerization rate, and the morphology and other properties of the final PU. In a more severe case, surfactant-catalyst interactions may lead to a reduction in their individual efficiencies or, in extreme scenarios, complete deactivation. This can result in incomplete or failed reactions, ultimately leading to lower-quality products.

To improve the accuracy of the models, these assumptions can be reinforced by detailed understanding of and accounting for all possible reactions including their interactions, and by more information on material character or understanding thereof. The approaches integrate phenomena at different levels (molecular, microstructural, and macroscopic) to provide a more comprehensive picture of the RPUF formation process. For example, and as mentioned earlier, accounting for the effect of the dynamic changes in the spatial freedom of reacting groups as the reaction proceeds, and surfactant effect and interactions can minimize errors and lead to more accurate predictions of the output material's final properties and behavior, leading to better quality products and more cost-efficient processes in their production. In this regard, designing targeted screening experiments that elucidate and quantify their effects and integrating them into computational models is crucial.

3.4.2 *Heuristics Versus Detailed Models*

The decision to use heuristics or detailed models for studying RPUFs hinges on various factors. Heuristics, while computationally less demanding, may sacrifice accuracy. They are well-suited for initial feasibility studies or when a general understanding of RPUF behavior is sufficient. Conversely, detailed models, though computationally more intensive, offer greater precision. They are indispensable for simulating complex RPUF formulations or predicting their behavior under extreme conditions and constraints.

Many studies have demonstrated the effectiveness of heuristics in modeling PU foam formation [23, 38, 41–48, 51–53]. Notably, Zhao et al. successfully simulated the RPUF formation of single polyol and multipolyol systems using heuristics outlined in Table 1, assuming catalyst effects on the gelling reaction but neglecting polyol-polyol interactions [8]. In subsequent research, they expanded their simulation framework to include catalyst effects on both gelling and blowing reactions [34], as well as varying catalyst types (organic and metallic) [23]. Building upon these findings, Al-Moameri et al. successfully simulated the RPUF formation of soy-based polyols using similar heuristics [54]. They further extended their simulations to investigate the effects of different physical blowing agents on polymerization temperature and height profiles [28, 55], and the influence of surfactants [51]. Ghoreshi et al. modeled RPUF formation by adding considerations on primary, secondary, and tertiary hydroxyl moieties [14] to existing heuristics, while Mendija et al. incorporated intrinsic amine in polyols into their simulation framework [38]. Alfeche et al. employed computational simulations to successfully improve the physico-mechanical properties of coconut oil-based RPUF following the heuristics in Table 1 with added assumptions.

References

1. R.G. Dingcong, A.A. Lubguban, L.C. Mendija, F.L. Alfeche, Compressive behavior and structural optimization of rigid polyurethane foams. Mater. Sci. J. (2023)
2. F.L.A. Alfeche, R.G. Dingcong, L.C. Mendija, A.A. Lubguban, In silico investigation of the impact of reaction kinetics on coconut-oil-based rigid polyurethane foam properties. Sustainability 15(9), 7148 (2023)
3. D. Whisler, H. Kim, Experimental and simulated high strain dynamic loading of polyurethane foam. Polym. Testing 41, 219–230 (2015). https://doi.org/10.1016/j.polymertesting.2014.12.004
4. H. Wang, Y. Liu, L. Lin, Behavior characteristics and thermal energy absorption of physical blowing agents in polyurethane foaming process. Polymers 15, 2285 (2023)
5. H.H. Al-Moameri, L.A. Jaf, G.J. Suppes, Simulation approach to learning polymer science. J. Chem. Educ. (2018)
6. C. Yang, L. Fischer, S. Maranda, J. Worlitschek, Rigid polyurethane foams incorporated with phase change materials: a state-of-the-art review and future research pathways. Energy Buildings 105, 164–178 (2015)
7. N. Usta, Investigation of fire behavior of rigid polyurethane foams containing fly ash. J. Appl. Polym. Sci. 124, 3372–3382 (2011)
8. Y. Zhao, M.J. Gordon, A. Tekeei, F.-H. Hsieh, G.J. Suppes, Modeling reaction kinetics of rigid polyurethane foaming process. J. Appl. Polym. Sci. 130(4), 1131–1138 (2013)
9. C.J. Permann, MOOSE: enabling massively parallel multiphysics simulation. SoftwareX 11, 100430 (2020)
10. C. Zhang et al., The foaming dynamic characteristics of polyurethane foam. J. Cell. Plast. 56(3), 279–295 (2019)

11. R.G. Dingcong et al., An iterative method for the simulation of rice straw-based polyol hydroxyl moieties. Sustainability **15**(15), 12082 (2023)

12. L. Ugarte, S. Gómez-Fernández, C. Peña-Rodríuez, A. Prociak, M.A. Corcuera, A. Eceiza, Tailoring mechanical properties of rigid polyurethane foams by sorbitol and corn derived Biopolyol mixtures. ACS Sustain. Chem. Eng. **3**(12), 3382–3387 (2015)

13. S. Miao, S. Zhang, Z. Su, P. Wang, Synthesis of bio-based polyurethanes from epoxidized soybean oil and isopropanolamine. J. Appl. Polym. Sci. **127**(3), 1929–1936 (2012)

14. R. Ghoreishi, Y. Zhao, G.J. Suppes, Reaction modeling of urethane polyols using fraction primary secondary and hindered-secondary hydroxyl content. J. Appl. Polym. Sci. **131**(12) (2014)

15. M.F. de Souza, J. Choi, S. Bhoyate, P.K. Kahol, R.K. Gupta, Expendable graphite as an efficient flame-retardant for novel partial bio-based rigid polyurethane foams. C **6**(2), 27 (2020)

16. C. Peptu, A.-D. Diaconu, M. Danu, C.A. Peptu, M. Cristea, V. Harabagiu, The influence of the hydroxyl type on crosslinking process in cyclodextrin based polyurethane networks. Gels **8**(6), 348 (2022)

17. T.N. Tuan Ismail et al., Urethane-forming reaction kinetics and catalysis of model Palm Olein polyols: quantified impact of primary and secondary hydroxyls. J. Appl. Polym. Sci. **133**(5) (2015)

18. M. Zieglowski et al., Reactivity of isocyanate-functionalized lignins: a key factor for the preparation of lignin-based polyurethanes. Front. Chem. **7** (2019)

19. D. Grzęda, G. Węgrzyk, A. Nowak, J. Idaszek, L. Szczepkowski, J. Ryszkowska, Cytotoxic properties of polyurethane foams for biomedical applications as a function of isocyanate index. Polymers **15**(12), 2754 (2023)

20. J. Zou, Y. Chen, M. Liang, H. Zou, Effect of hard segments on the thermal and mechanical properties of water blown semi-rigid polyurethane foams. J. Polym. Res. **22**(6) (2015)

21. T. Nagy et al., New insight into the kinetics of diisocyanate-alcohol reactions by high-performance liquid chromatography and mass spectrometry. J. Appl. Polym. Sci. **132**(25) (2015)

22. H. Ni, H.A. Nash, J.G. Worden, M.D. Soucek, Effect of catalysts on the reaction of an aliphatic isocyanate and water. J. Polym. Sci., Part A: Polym. Chem. **40**(11), 1677–1688 (2002)

23. J. Li, S. Jiang, L. Ding, L. Wang, Reaction kinetics and properties of MDI base poly (urethane-isocyanurate) network polymers. Des. Monomers Polym. **24**(1), 265–273 (2021)

24. Y. Zhao, G.J. Suppes, Computational study on reaction enthalpies of urethane-forming reactions. Polym. Eng. Sci. **55**(6), 1420–1428 (2015)

25. A.S. Dutta, Polyurethane foam chemistry. Recycling of polyurethane foams, 17–27 (2018)

26. R. Tesser, M. Serio, A. Sclafani, E. Santacesaria, Modeling of polyurethane foam formation. J. Appl. Polym. Sci. **92**(3), 1875–1886 (2004)

27. L. Jaf, H.H. Al-Moameri, A.A. Ayash, A.A. Lubguban, R.M. Malaluan, T. Ghosh, Limits of performance of polyurethane blowing agents. Sustainability **15**(8), 6737 (2023)

28. H. Al-Moameri, Y. Zhao, R. Ghoreishi, G.J. Suppes, Simulation of liquid physical blowing agents for forming rigid urethane foams. J. Appl. Polym. Sci. **132**(34) (2015)

29. H.H. Al-Moameri, G. Hassan, B. Jaber, Simulation physical and chemical blowing agents for polyurethane foam production. IOP Conf. Series: Mater. Sci. Eng. **518**(6), 062001 (2019)

30. A.A. Mahmoud, E.A. Nasr, A.A. Maamoun, The influence of polyurethane foam on the insulation characteristics of mortar pastes. J. Miner. Mater. Charact. Eng. **05**(02), 49–61 (2017)

31. J. Liu, Z. Sun, F. Wang, D. Zhu, J. Ge, H. Su, Facile solvent-free preparation of biobased rigid polyurethane foam from raw citric acid fermentation waste. Ind Eng. Chem. Res. **59**(22), 10308–10314 (2020)

32. S. Wang, H. Chen, L. Zhang, Thermal decomposition kinetics of rigid polyurethane foam and ignition risk by a hot particle. J. Appl. Polym. Sci. **131**(4) (2013)

33. J.C.S. Bondaug et al., Development of a catalyst system for enhanced properties of coconut diethanolamide-based rigid poly(urethane-urea) foam. ACS Appl. Polym. Mater. **6**(11), Am. Chem. Soc. (ACS), 6875–6887 (2024). https://doi.org/10.1021/acsapm.4c01187

34. Y. Zhao, F. Zhong, A. Tekeei, G.J. Suppes, Modeling impact of catalyst loading on polyurethane foam polymerization. Appl. Catal. A **469**, 229–238 (2014)
35. R.M. Hill, Silicone surfactants—new developments. Curr. Opin. Colloid Interface Sci. **7**(5–6), 255–261 (2002)
36. Y. Zhao, H. Al-Moameri, R. Ghoreishi, G. Suppes, Modeling impact of surfactants on polyurethane foam polymerization (2015)
37. P. Sharma, V.R. Prasath Kumar, Prediction and analysis of sandwich panel with rice husk and polyurethane foam using machine learning model. Asian J. Civil Eng. **24**(8), 3009–3021 (2023)
38. L.C.C. Mendija, R.G. Dingcong Jr., F.L.A.M. Alfeche, R.M. Malaluan, A.C. Alguno, A.A. Lubguban, Elucidating the impact of polyol functional moieties on exothermic poly(urethane-urea) polymerization: a thermo-kinetic simulation approach. Sustainability **16**(11), 4587 (2024)
39. Dow, Flexible PU foams, Second. The Dow Chemical Company (1997)
40. M. Ionescu, *Chemistry and Technology of Polyols for Polyurethanes*. Rapra Technology Ltd, Shropshire, UK (2005)
41. P. Krol, Synthesis methods, chemical structures and phase structures of linear polyurethanes. Properties and applications of linear polyurethanes in polyurethane elastomers, copolymers and ionomers Prog. Mater. Sci. **52**, 915–1015 (2007)
42. Y. Zhao, *Modeling and Experimental Study of Polyurethane Foaming Reactions* (University of Missouri-Columbia, Columbia, MO, USA, 2015)
43. H.H. Al-Moameri, *Simulation Aided Analysis and Experimental Study of Polyurethane Polymerization Reaction and Foaming Process* (University of Missouri-Columbia, Columbia, MO, USA, 2017)
44. L. Jaf, *Modeling and Experimental Study of Polyurethane Foaming and Gelling Reactions* (University of Missouri-Columbia, Columbia, MO, USA, 2018)
45. H. Rusche, M. Karimi, P. Ferkl, S. Karolius, Simulating polyurethane foams using the MoDeNa multi-scale simulation framework. OpenFOAM® 61, 401–417 (2019)
46. C. Raimbault et al., Foaming parameter identification of polyurethane using FOAMAT® device. Polym. Eng. Sci. **61**, 1243–1256 (2021)
47. H.C. Wright, D.D. Cameron, A.J. Ryan, FoamPi: an open-source raspberry Pi-based apparatus for monitoring polyurethane foam reactions. HardwareX **12**, 00365 (2022)
48. X. Leng et al., A study on coconut fatty acid diethanolamide-based polyurethane foams. RSC Adv. **12**, 13548–13556 (2022)
49. R. Dingcong et al., A novel reaction mechanism for the synthesis of coconut oil-derived biopolyol for rigid poly(urethane-urea) hybrid foam application. RSC Adv. **13**, 1985–1994 (2023)
50. S. . Baser, D.V.P. Khakhar, Eng. Sci. **34**, 642 (1994)
51. H. Al-Moameri, Y. Zhao, R. Ghoreishi, G. Suppes, Simulation silicon surfactant rule on polyurethane foaming reactions. Iran. J. Chem. Chem. Eng. **40**, 1256–1268 (2021)
52. P. Ferkl, I. Kršková, J. Kosek, Evolution of mass distribution in walls of rigid polyurethane foams. Chem. Eng. Sci. **176**, 50–58 (2018)
53. D. Hu, C. Zhou, T. Liu, Y. Chen, Z. Liu, L. Zhao, Experimental and numerical study of the polyurethane foaming process using high-pressure CO_2. J. Cell. Plast. **57**, 927–949 (2021)
54. H. Al-Moameri, R. Ghoreishi, Y. Zhao, G.J. Suppes, Impact of the maximum foam reaction temperature on reducing foam shrinkage. RSC Adv. **5**(22), 17171–17178 (2015)
55. H. Al-Moameri, Y. Zhao, R. Ghoreishi, G.J. Suppes, Simulation blowing agent performance, cell morphology, and cell pressure in rigid polyurethane foams. Ind. Eng. Chem. Res. **55**(8), 2336–2344 (2016)
56. L. Shen, Y. Zhao, A. Tekeei, F.H. Hsieh, G.J. Suppes, Density modeling of polyurethane box foam. Polym. Eng. Sci. **54**(7), 1503–1511 (2014)
57. M. Sadouki, R. Bengherbia, A. Mahiou, Measuring audible acoustic frequencies parameters of rigid porous media via an innovative impedance tube method-Solving the inverse problem. INTER-NOISE NOISE-CON Congress Conf. Proc. **270**(11), 170–177 (2024)
58. Z. Hu, J. Shao, S. Jia, W. Song, C. Wang, Propagation properties of shock waves in polyurethane foam based on atomistic simulations. Def. Technol. **31**, 117–129 (2024)

Chapter 4
Implications and Future Outcomes

4.1 Predictive Formulation Techniques

Innovations in Formulation Prediction. Predictive formulation has evolved into a cornerstone of modern RPUF research, bringing precision and efficiency to a field traditionally dominated by empirical trial and error. Machine learning (ML) algorithms now analyze vast datasets of chemical compositions, mechanical properties, and environmental influences, delivering a level of predictive power previously unavailable. Unlike conventional methods that rely on repetitive testing, ML models draw from extensive data to make accurate predictions about RPUF properties, enabling researchers to optimize formulations to a molecular level. For instance, advanced algorithms like neural networks can determine optimal polyol-isocyanate ratios, which directly influence foam properties such as thermal insulation, compressive strength, and fire resistance [1]. By enabling the fine-tuning of these properties, predictive models help researchers develop foams for specialized applications, such as high-performance thermal insulation in buildings or lightweight materials for automotive parts. Such applications have already demonstrated performance improvements of 10–20% over traditionally formulated foams, underscoring the effectiveness of ML-enhanced formulation techniques.

Machine learning has significantly expanded the predictive capabilities of RPUF formulation by enabling the use of more complex algorithms, such as convolutional neural networks and gradient-boosted decision trees. These models analyze multidimensional data, identifying intricate relationships among formulation variables that affect end-product properties. This complexity allows researchers to create tailored RPUFs that meet the specific demands of diverse industries, such as construction, where improved thermal resistance can lead to significant energy savings, or in the automotive sector, where lightweight foams reduce fuel consumption without compromising safety. Recent applications in insulation products, for example, have achieved a 15–20% reduction in thermal conductivity using ML-driven formulation adjustments, setting a new industry benchmark for energy-efficient materials

A. A. Lubguban et al., *Computational Thermo-kinetics of Rigid Polyurethane Foams*, SpringerBriefs in Applied Sciences and Technology, https://doi.org/10.1007/978-981-96-2077-7_4

[2]. As ML technologies continue to advance, RPUF research will likely evolve from a predominantly experimental science into a computationally driven field, with ML models guiding nearly every stage of product development from raw material selection to final quality testing.

In spite of the tremendous potential of predictive formulation, data quality and integration remain critical challenges in leveraging ML to its fullest potential in RPUF research. Data consistency is crucial, as inaccuracies in experimental data or variations in dataset structure can undermine the predictive accuracy of ML models. Effective data preprocessing techniques, such as normalization and noise reduction, are essential to enhance data quality and improve the reliability of model predictions [3]. Furthermore, innovative data augmentation approaches, like synthetic data generation, have allowed researchers to overcome limitations in dataset size, enhancing model robustness. As the field progresses, efforts to standardize data collection and reporting practices across institutions and industries will play a critical role in ensuring that predictive models are built on high-quality, reliable data. With standardized data and advanced ML algorithms, RPUF researchers can unlock new levels of accuracy and efficiency, positioning predictive formulation as a cornerstone in the pursuit of sustainable, high-performance foam materials.

Optimization and Efficiency. Optimization and efficiency gains achieved through computational modeling in RPUF formulation have redefined industry standards, reducing resource use, time-to-market, and environmental impact. Traditionally, RPUF development required extensive experimental testing, an often laborious process that consumed significant quantities of raw materials, energy, and time. By simulating diverse formulations and testing parameters, computational models eliminate much of the trial and error, enabling researchers to rapidly refine formulations and reach optimal configurations. This streamlined approach has reduced development times by up to 60% for some companies, with accompanying reductions in production costs of up to 40% [4]. As industries face increasing pressure to enhance efficiency and sustainability, computational optimization provides a powerful tool for reducing resource consumption without compromising product quality. Computational models allow for the precise adjustment of material proportions, achieving optimal performance while minimizing waste, a critical advantage in an era of stringent environmental regulations and resource scarcity.

Resource efficiency is a primary advantage of predictive modeling, especially in reducing reliance on nonrenewable petrochemicals. By predicting the performance of alternative materials, such as bio-based polyols, computational models offer environmentally conscious alternatives that do not sacrifice durability or insulation capacity. For instance, a recent study indicated that bio-based polyol formulations could reduce the carbon footprint of RPUF production by approximately 30%, aligning with global sustainability goals and helping companies meet increasingly rigorous emissions standards [5]. Computational models enable these advancements by allowing researchers to explore and evaluate different formulations under simulated conditions, avoiding the costs and environmental impact associated with extensive laboratory trials. This capacity for virtual experimentation allows manufacturers to innovate sustainably, balancing economic and ecological goals while meeting consumer demands for greener products. As the RPUF industry continues to adopt these methods, the role of predictive models in promoting sustainable production practices is likely to expand.

Energy efficiency in production is another significant benefit derived from predictive models, particularly in the optimization of foam expansion and curing processes. These computational models can fine-tune parameters that influence energy consumption, such as curing temperature and pressure, reducing energy usage by up to 30% in some cases. Such savings not only lower production costs but also contribute to environmental sustainability, as reduced energy demand translates to fewer carbon emissions. Additionally, predictive models facilitate material recycling and the reintegration of production waste, promoting a circular economy approach in RPUF manufacturing. Companies that have implemented closed-loop manufacturing systems report cost savings of up to 25% and reductions in waste generation by nearly half, demonstrating the economic viability of resource-efficient production [6]. As computational modeling becomes integral to RPUF manufacturing, these efficiencies are expected to set new industry standards, particularly as consumer and regulatory pressures increase the demand for sustainable production practices.

Case Studies and Commercial Impact. Predictive modeling has had a measurable impact on the commercialization of high-performance RPUFs, helping companies develop products that meet stringent industry standards and exceed traditional performance metrics. In the automotive industry, for example, RPUFs optimized through ML-driven formulations are increasingly used to reduce vehicle weight, contributing to improved fuel efficiency and lower emissions. Some of these automotive foams are up to 15% lighter than conventional alternatives while maintaining comparable strength and durability, underscoring the role of predictive models in enabling manufacturers to meet fuel efficiency and sustainability targets [7]. This weight reduction is particularly valuable in electric vehicles, where lighter materials contribute to extended battery life, further promoting the transition to sustainable transportation solutions. In addition to vehicle manufacturing, the construction industry has benefited significantly from computationally optimized RPUFs, particularly in energy-efficient building insulation. Predictive models have allowed manufacturers to develop insulation foams with enhanced thermal resistance, which can reduce building energy consumption for heating and cooling by up to 25%. These advanced insulation products not only lower energy costs for consumers but also align with global initiatives to improve energy efficiency in the built environment [8].

The application of FAIR principles (Findable, Accessible, Interoperable, and Reusable) has further advanced the commercial potential of RPUFs by fostering a collaborative research environment. Open-access data initiatives based on FAIR principles enable a wider range of researchers and manufacturers to access valuable datasets, facilitating innovation and reducing redundancy in RPUF research and development. By sharing non-proprietary data, companies can contribute to a growing knowledge base that benefits the industry as a whole, while still protecting intellectual property related to proprietary formulations. This open-data approach has proven especially beneficial for smaller companies, which can leverage high-quality datasets to accelerate their own product development, leveling the playing field within the RPUF market and accelerating industry-wide advancement [9]. The adoption of FAIR principles is expected to continue shaping the RPUF sector, enabling faster and more cost-effective innovation while supporting a collaborative research ecosystem that drives sustainable development.

4.2 Future Directions for Computational Modeling

Emerging Trends. The field of RPUF computational modeling is advancing rapidly, and emerging trends are reshaping how researchers approach RPUF development. One of the most promising directions is the adoption of hybrid modeling approaches, which combine traditional computational techniques with newer data-driven methods. Hybrid models integrate the strengths of empirical models with machine learning-based predictions, allowing for enhanced accuracy and flexibility in RPUF simulations. For example, traditional Finite Element Analysis (FEA) provides a strong foundation for understanding mechanical behavior under stress, while machine learning can predict outcomes based on complex patterns in historical data. Together, these approaches have improved the precision of RPUF simulations, particularly in scenarios involving complex, interacting phenomena such as temperature gradients, mechanical stress, and chemical reactions [10]. By creating hybrid models, researchers can simulate RPUF behaviors that are otherwise challenging to predict, such as combined thermal and mechanical degradation, improving the reliability of these models in real-world applications [4].

In addition to hybrid modeling, AI-driven methods are increasingly influential in RPUF research. Artificial intelligence, particularly through machine learning and deep learning, can analyze large datasets to identify optimal formulations and predict foam performance with high accuracy. AI techniques, such as neural networks, can capture nonlinear relationships between variables, which are essential for complex materials like RPUFs where multiple interacting components affect overall performance. For instance, AI has been used to develop new polyol blends that enhance foam properties, such as thermal resistance and compressive strength, by predicting how different formulations will perform under specific conditions. Researchers have successfully applied AI to design foams with tailored characteristics for specialized applications, such as ultra-lightweight RPUFs for the aerospace industry or RPUFs with improved insulation properties for construction [9]. As AI-driven approaches continue to advance, they are expected to play an increasingly critical role in discovering new material formulations and optimizing RPUF production processes.

Future Applications. The future of RPUF computational modeling holds considerable potential, with emerging technologies poised to address some of the most complex challenges in the field. One such frontier is quantum computing, which has the potential to solve highly complex material science problems that are currently beyond the reach of classical computers. Quantum computing could enable faster and more precise simulations of molecular interactions within RPUFs, providing deeper insights into how foam structures respond to various environmental factors. Quantum algorithms, designed to handle enormous datasets and complex interactions, could revolutionize RPUF modeling by predicting properties at the atomic and molecular levels, which is critical for designing foams with superior mechanical and thermal properties [8].

Another area with transformative potential is the integration of big data analytics. As experimental data on RPUF formulations and performance grows, big data tools are becoming essential for analyzing this information and extracting valuable insights. Big data analytics can handle vast amounts of diverse data, allowing researchers to uncover correlations and trends that were previously difficult to detect. For example, by analyzing data from thousands of RPUF samples, researchers can identify the factors that most significantly impact foam performance and sustainability. This information can then inform the development of RPUFs that are both high-performing and environmentally friendly [11]. Combined with machine learning, big data analytics is enabling more accurate predictions, faster computation times, and the ability to model more complex RPUF systems.

Additionally, the development of new algorithms for multiscale modeling is advancing the ability to simulate RPUF behavior across various scales, from molecular interactions to macroscopic foam properties. Multiscale modeling allows researchers to link material behavior at the microscopic level to observable characteristics at the product level, improving the understanding of how molecular structures influence bulk properties such as durability and thermal conductivity [12]. By bridging the gap between micro- and macrolevels, multiscale modeling can provide a more comprehensive view of RPUF performance, allowing for targeted improvements in formulation and design. These developments promise to significantly enhance the accuracy and applicability of RPUF models, providing a foundation for the next generation of high-performance, sustainable foams.

Interdisciplinary Collaboration. The future of RPUF computational modeling relies not only on technological advances but also on interdisciplinary collaboration. As RPUF modeling becomes more complex, the integration of expertise from various fields—including material science, chemical engineering, computer science, and data analytics—has become essential. Material scientists bring a deep understanding of RPUF chemistry and behavior, while chemical engineers provide insights into processing conditions and scalability. Computer scientists and data analysts, on the other hand, contribute expertise in algorithms, data processing, and model development, ensuring that RPUF simulations are both accurate and computationally efficient [13].

Collaborative efforts across these disciplines are already leading to more robust RPUF models that can address the multifaceted challenges of foam design and application. For instance, joint projects between material scientists and data scientists have led to the development of machine learning algorithms that improve model accuracy by incorporating chemical and physical principles directly into the computational framework. These collaborations have been instrumental in creating predictive models capable of handling the complexities of RPUF formulations, such as the interaction between various polyols and catalysts under specific environmental conditions [3]. As these interdisciplinary partnerships continue to grow, they are expected to drive innovation in RPUF research, leading to the development of more sophisticated models and formulations that meet the evolving demands of industries ranging from automotive to construction.

The importance of interdisciplinary collaboration is further underscored by the need to address sustainability challenges in RPUF production. As regulatory pressures and consumer expectations for eco-friendly products increase, researchers from different fields must work together to develop RPUFs that are both high-performing and environmentally sustainable. Material scientists, for instance, are exploring bio-based alternatives to traditional petrochemical components, while data scientists are applying big data analytics to assess the environmental impact of different formulations over their lifecycles. By combining these efforts, interdisciplinary teams can create RPUFs that meet stringent environmental standards without sacrificing performance [14]. Through collaborative innovation, RPUF computational modeling can continue to evolve, contributing to a sustainable future and expanding the possibilities for foam technology in various applications.

4.2.1 Application to Academia and Industry

Advancements in computational modeling are transforming both academic and industrial approaches to researching and developing rigid polyurethane foams (RPUFs). These models enable detailed explorations of the RPUF properties by researchers as well as industry professionals such that the real-world operating conditions can be simulated along with predictions of RPUF behavior across various applications.

In *academic settings*, computational modeling serves as a powerful tool for researchers to conduct in-depth studies on rigid polyurethane foams (RPUFs). These models facilitate the simulation of mechanical, thermal, and chemical properties of RPUFs, which can yield insights that are often unattainable through traditional experimental methods alone. For instance, He et al. proved that using computational analysis could be an efficient way of understanding molecular structure and mechanical properties for thermally treated RPUF which showed that chemical reactivity exhibited significant variations as temperatures rise [15]. This aligns with findings from Wang et al., who emphasized the importance of cell morphology in determining the thermal insulation and mechanical properties of RPUFs, suggesting that computational modeling can help optimize these characteristics by simulating different structural configurations [16].

Moreover, the ability to simulate various conditions allows researchers to explore the effects of different additives and fillers on RPUFs without the constraints of physical experimentation. For example, Akdoğan et al. highlighted how the incorporation of flame retardants can influence the thermal insulation and mechanical properties of RPUFs, a process that can be effectively modeled to predict outcomes before actual implementation [17]. According to Delucis et al., computational modeling

can even account for the interactions of the fillers with the RPUF matrix, important in terms of improving mechanical and thermal performance [18]. Simulating
complex interactions may thus save time and resources, but it also enables the possibility of innovative experiments that otherwise would be limited due to physical and
financial constraints.

Furthermore, through computational modeling coupled with experimental validation, more complete insights into the behavior of RPUF under a variety of conditions
can be achieved. Wu et al. used the finite element method to predict the elasto-
viscoplastic behavior of RPUFs, thus proving the advantages achieved by combining
computation with empirical data in order to generate more reliable predictions [19].
This synergy of modeling and experimentation is quite essential to propel the advance
in materials science, especially to optimize RPUF for a myriad of applications across
construction, refrigeration, and much more [20].

In material science, computational modeling offers an unprecedented opportunity
to explore the molecular structure and behavior of rigid polyurethane foams, which
is bound to provide fundamental insights into their properties. Through examination
of molecular arrangements and microstructural configurations, researchers are able
to determine the mechanisms that give RPUFs unique qualities such as thermal
insulation, compressive strength, and durability. For example, Kamairudin et al.
pointed out that the properties of RPUFs can be significantly influenced by the
molecular structure of the polyols used, which directly affects thermal insulation
capabilities and mechanical performance [21]. This assertion is supported by Adnan
et al., who noted that the compressive strength of RPUFs is closely related to their
density and microstructural characteristics, which can be effectively modeled to
predict performance outcomes [22].

Additionally, the modeling capability allows researchers to simulate interactions in the polymer matrix and enlighten how specific configurations might impact
material performance. Kabakci et al. emphasized the importance of establishing
processing-structure–property relationships in RPUFs by pointing out that differences in cell morphology can lead to significant differences in thermal and mechanical properties [23]. The further emphasis on microstructural analysis is by Acuña
et al. in an explanation of the microstructural properties of the fillers in a polymer
matrix in affecting the mechanical strength and thermal stability of RPUFs, thereby
highly emphasizing critical importance in their optimization material performance
[24].

Moreover, such interactions have been simulated in order to enhance understanding of RPUF behavior, as well as provide capability for new formulations
that might be designed to meet application-specific requirements. For instance, the
authors Chen et al. established the effects of expansionable graphite on morphology
as well as thermal properties of RPUFs. This in effect showed that the dispersion can
offer a great deal regarding improvements in flame retardants yet maintain desirable
mechanical characterization [25]. This modeling approach allows for the identification of optimal filler types and concentrations, which is crucial for tailoring RPUFs
for various applications in the construction and insulation industries.

A major focus of computational modeling in the study of rigid polyurethane foams (RPUFs) involves simulating the molecular chains within the polyurethane structure, which consist of hard and soft segments that contribute differently to material properties. Hard segments provide rigidity and structural integrity, while soft segments impart flexibility and resilience. Oprea et al. proved that the hard-segment structure plays a significant role in determining the mechanical properties of polyurethane elastomers: changes in the mobility of chains of the hard segments influence the overall stiffness/flexibility of the material [26]. This modeling capability allows researchers to test various arrangements and ratios of these segments, enhancing the understanding of how RPUFs might behave under different conditions, such as extreme temperatures or high-stress environments.

The computational models allow researchers to simulate the thermal conductivity of rigid polyurethane foams at the molecular level. It studies how heat transfers through both solid and gas phases within foam cells. For example, Akdoğan et al. highlighted that the thermal conductivity of RPUFs is affected by mean cell size, density, and the thermal conductivity of the blowing agent within the cells [17]. Analysis of gas behavior within foam cells can be used to optimize foam densities and pore structures to maximize insulation, hence lighter and more efficient solutions. This optimization is particularly relevant in applications where thermal performance is essential, such as in building insulation and vehicle components.

In addition to thermal performance, computational models facilitate studies on the resilience of RPUFs under mechanical stress. By simulating stresses at a microstructural level, researchers can predict how RPUFs might deform or fail under specific loads. As Sang et al. noted, it is the soft segment up to the degree of microphase separation and that of the hard segment which determine the mechanical properties of the polyurethane elastomer, and this in turn determines their performance under stresses [27]. It even predicts their potential applications wherein impact resistance is a concern factor, such as those used in automotive interiors and construction panels. Researchers refine polymer composition with additives or crosslinking agents to improve material strength and toughness so that the RPUF can meet very demanding specifications [28].

In industry, computational modeling serves as a bridge between academic research and practical applications. By providing predictive data, computational models enable companies to accelerate product development, moving more quickly from design to production with reduced reliance on costly physical testing. This modeling capability is transformative across multiple sectors.

In the automotive industry, rigid polyurethane foams (RPUFs) are utilized for thermal insulation, impact absorption, and lightweight, which are essential for fuel efficiency and safety. Computational models allow automotive engineers to simulate the thermal and impact performance of RPUFs, predicting how they will function as insulation within engine compartments or as impact-absorbing padding in vehicle interiors. Wu et al. highlighted the advantages of RPUFs in automotive applications, noting their low density and high insulation performance, which contribute to vehicle efficiency and safety [19]. Engineers can fine-tune properties like material composition, cell size, and density, ensuring optimal insulation and impact absorption.

In the construction industry, RPUFs are critical for energy-efficient building insulation. Computational models allow architects and engineers to simulate RPUF thermal performance across various climate scenarios, determining ideal foam density and cell structure to optimize insulation. Magiera emphasized the favorable properties of RPUFs, such as low apparent density and high resistance to heat, making them suitable for insulation purposes in construction [29]. Models also allow for modifications tailored to specific applications, such as roof versus wall insulation, which have different thermal and structural requirements. Given the global push for sustainability, these models help ensure compliance with stringent building codes focused on energy efficiency.

In marine engineering, RPUFs are valued for their lightweight buoyancy and moisture resistance, making them ideal for applications like boat hulls and flotation devices. Computational models enable engineers to test RPUF durability and stability in virtual marine environments, accounting for prolonged exposure to saltwater, temperature variations, and mechanical wear. Hoseinabadi et al. discussed the synthesis of RPUFs with nanoporous graphene, which enhances their mechanical and thermal properties, thus improving their performance in marine applications [30]. Engineers can experiment with additives or alternative formulations that enhance moisture resistance, improving product longevity and reducing maintenance costs.

Beyond performance optimization, computational modeling assists industries in meeting regulatory standards while refining production processes. By simulating performance at each stage, these models allow manufacturers to streamline production, minimizing waste and enhancing quality control. Hürkamp et al. noted the importance of integrating computational concepts in product and production engineering to support concurrent engineering and improve manufacturing efficiency [31].

4.2.2 Regulatory and Standardization Efforts

Computational modeling is a critical tool in supporting regulatory compliance and establishing industry standards for RPUFs, particularly in terms of performance, safety, and environmental impact. With increasing regulatory requirements at both regional and international levels, computational models have become essential for predicting RPUF performance and ensuring compliance with safety and environmental standards before physical testing.

Safety Standards. A primary regulatory concern with rigid polyurethane foams (RPUFs) is fire safety, especially since they are widely used in building insulation and automotive interiors. Meeting fire safety standards, such as those mandated in ASTM E84 for flame spread, is crucial. Computational models allow manufacturers to simulate fire exposure and predict how different formulations respond to flames.

A demonstration by Zhang et al. shows that incorporation of intumescent flame-retardant coatings significantly improves the flame-retardant performance of RPUFs, allowing manufacturers to assess the effects on flame spread, smoke production, and toxicity [32]. This predictive modeling helps companies ensure compliance with fire safety requirements before conducting physical tests, reducing the risk of recalls and enhancing product safety.

Performance Standards. Beyond safety, RPUFs must meet performance standards related to thermal insulation and load-bearing capacity. For example, RPUFs used in construction are expected to maintain thermal resistance (R-value) over extended periods in different environments. Singh emphasized that RPUFs exhibit excellent thermal insulation properties, which can be optimized through computational modeling to simulate external factors like temperature, humidity, and UV exposure, predicting potential changes in thermal resistance [33]. Similarly, in automotive and aerospace applications, models simulate mechanical stresses that RPUFs will encounter, ensuring they meet impact resistance and durability requirements in demanding conditions. By testing these performance standards in virtual environments, manufacturers can have confidence in product reliability before physical trials.

Environmental Standards. RPUFs' environmental impact is increasingly scrutinized as industries prioritize sustainability. Regulatory bodies are beginning to require data on RPUF recyclability, emissions, and energy efficiency over the product lifecycle. He et al. discussed the importance of recycling waste polyurethane foams, highlighting how computational models can quantify the carbon footprint reduction of RPUFs used in insulation or lightweight applications [34]. This supports compliance with standards like LEED certification for buildings and Environmental Product Declarations (EPDs) in construction. A concrete example of computational modeling driving environmental standards is in the automotive industry, where RPUFs' lightweight capability helps reduce vehicle emissions. Computational models can predict the effect of RPUFs on fuel consumption by simulating their impact on vehicle weight, helping manufacturers meet emissions regulations like the EU's CO_2 targets or the U.S. CAFE standards.

In construction, models help set standards for energy-efficient RPUFs by evaluating their thermal conductivity in various climates. Dunn et al. emphasized that understanding the lifecycle energy consumption and emissions of materials, including RPUFs, is crucial for meeting environmental standards [35]. This comprehensive approach ensures that RPUFs not only perform well but also align with sustainability goals.

As the industry evolves, computational models hold promise for developing new regulatory frameworks, particularly for sustainable and bio-based RPUFs. By simulating bio-based RPUF properties, manufacturers can validate environmental claims and demonstrate compliance with green materials standards. This data-driven approach also supports the development of benchmarks for biodegradable RPUFs, accelerating the adoption of sustainable materials across sectors.

4.2.3 Sustainability and Environmental Considerations

One of the most transformative roles of computational modeling in the RPUF industry is its contribution to sustainability and environmental responsibility. These models allow companies to reduce energy consumption, minimize waste, and explore renewable resources, supporting global sustainability goals.

Enhancing Production Efficiency and Reducing Waste. Computational models help manufacturers predict the energy demands of rigid polyurethane foam (RPUF) production and adjust materials or techniques accordingly. For instance, by simulating the impact of lower-temperature curing processes, manufacturers can reduce energy consumption without sacrificing product quality. Liu et al. demonstrated that optimizing the curing temperature in RPUF production can lead to significant energy savings while maintaining the mechanical properties of the foam [36]. Models can also assess closed-loop manufacturing, where materials are recycled within the production cycle, conserving resources and reducing waste.

Facilitating the Use of Renewable and Recycled Materials. As the industry transitions away from petrochemical-based foams, computational models offer safe, efficient ways to explore bio-based and recycled materials. For example, Zhang et al. explored the use of bio-based polyols derived from plant oils in RPUFs, indicating that these materials can enhance foam properties like rigidity and insulation [37]. If models indicate that these materials meet performance requirements, companies can proceed confidently with physical testing, reducing both financial and environmental risks. This modeling approach supports circular economy practices, extending material lifecycles and reducing reliance on nonrenewable resources.

Optimizing RPUFs for Energy Efficiency in Buildings. RPUFs are mostly utilized in infrastructures as insulation materials to lower heating and cooling demands. Computational models allow manufacturers to optimize the thermal properties of RPUFs, creating foams that maximize insulation while using minimal resources. For instance, Soykan and Kaya demonstrated that by simulating different percentages of hemp fiber in RPUFs, manufacturers can identify ideal structures that provide high insulation while reducing material use [38]. This optimization not only reduces material use but also supports sustainable building practices by lowering buildings' carbon footprints.

Contributing to Fuel Efficiency in the Automotive Industry. Lightweight RPUFs are essential in the automotive industry's push for fuel efficiency. By integrating RPUFs in vehicle interiors and structural parts, manufacturers reduce vehicle weight, which lowers fuel consumption. Computational models enable manufacturers to simulate RPUFs' effects on vehicle performance, including crashworthiness and insulation, ensuring that lightweight designs do not compromise safety standards. Wang et al. highlighted that the integration of RPUFs in automotive applications can significantly improve fuel efficiency while maintaining safety [39]. This modeling is increasingly valuable as emissions regulations tighten, helping companies comply with environmental standards while staying competitive.

Aligning with Global Environmental Goals. Optimized RPUFs contribute to broader environmental goals, such as those outlined in the Paris Agreement, by reducing greenhouse gas emissions. RPUFs that enhance building insulation or vehicle fuel efficiency directly reduce energy demands and emissions, supporting cleaner, more sustainable industry practices. As modeling advances, manufacturers may even simulate RPUFs' end-of-life scenarios, designing foams that prioritize recyclability or biodegradability, aligning with global waste reduction initiatives. Yang et al. discussed the potential of using bio-based flame retardants in RPUFs to enhance their environmental profile while maintaining performance [40].

Compliance and Public Perception Benefits. Companies aligning with sustainability through computational modeling benefit in regulatory compliance and public perception. Predictive models help companies proactively design RPUFs that meet standards like the European Green Deal. Additionally, eco-conscious consumers increasingly value companies committed to sustainability, enhancing brand reputation and market competitiveness. Akar et al. noted that companies that adopt sustainable practices and demonstrate compliance with environmental regulations can significantly improve their market position [41].

Overall, computational modeling is a vital force in promoting environmental responsibility within the RPUF industry. By simulating material properties, manufacturing processes, and environmental impacts, computational models empower companies to create sustainable products that conserve energy, minimize waste, and use renewable resources. The wide range of applications—from optimizing insulation in buildings to enhancing vehicle fuel efficiency—highlights the versatility of computational modeling in achieving sustainability goals. As the industry faces growing regulatory demands and consumer expectations for sustainable products, computational modeling will continue to drive responsible RPUF production, connecting academic innovation with practical, eco-conscious industrial applications.

References

1. R. Ramprasad, R. Batra, G. Pilania, A. Mannodi-Kanakkithodi, C. Kim, Machine learning in materials informatics: recent applications and prospects. NPJ Comput. Mater. **3**, 54 (2017)
2. J. Wei et al., Machine learning in materials science. InfoMat **1**(3), 338–358 (2019)
3. T. Lü, Y. Zhang, H. Yan, Z. Xu, Z. Liu, S. Han, Artificial intelligence assisted thermoelectric materials design and discovery. Materials **15**(1) (2022)
4. K.V. Lakshmi, Artificial intelligence and its applications in nanochemistry. Int. J. Eng. Technol. Manage. Sci. **7**(5), 385–389 (2023)
5. Z. Yang, Y. Wang, H. Luo, J. Li, M. Xu, Q. Wang, Development status and prospects of artificial intelligence in the field of energy conversion materials. Energy **202**, 117496 (2020)
6. N. Ravi, P. Chaturvedi, E.A. Huerta, Z. Liu, R. Chard, K.J. Schmidt, FAIR principles for AI models with a practical application for accelerated high energy diffraction microscopy. Sci. Data **9**(657), 1–12 (2022)
7. D. Drikakis, F. Sofos, Can artificial intelligence accelerate fluid mechanics research?. Comput Fluid. **222**, 104996 (2023)

8. A.K. Ligo, K. Rand, J. Bassett, I. Linkov, Comparing the emergence of technical and social sciences research in artificial intelligence. Front. Comput. Sci. **3**, 653235 (2021)

9. G. Chen, Q. Cheng, S. Puetz, Special issue: data-driven discovery in geosciences: opportunities and challenges. Math. Geosci. **55**(2), 287–293 (2023)

10. A. Karpatne et al., Theory-guided data science: a new paradigm for scientific discovery from data. IEEE Trans. Knowl. Data Eng. **29**(10), 2318–2331 (2017)

11. B. Blaiszik, R. Chard, L. Ward, I. Foster, A data ecosystem to support machine learning in materials science. MRS Bulletin **44**(6), 463–470 (2019)

12. M. Sorkun, A. Khetan, S. Maier, An artificial intelligence-aided virtual screening recipe for two-dimensional materials discovery. NPJ Comput. Mater. **6**, 63 (2020)

13. V. Marchenko, A. Alekseeva, S. Efimov, V. Korolev, A. Marchenko, A. Kuznetsov, Database of two-dimensional hybrid perovskite materials: open-access collection of crystal structure. NPJ Comput. Mater. **6**, 42 (2020)

14. B. Boehm, J. Lane, T. Hilburn, INTERSECT architecture specification system-of-systems architecture (2022)

15. Y. He, D. Qiu, Z. Yu, Multiscale investigation on molecular structure and mechanical properties of thermal-treated rigid polyurethane foam under high temperature. J. Appl. Polym. Sci. **138**(44) (2021)

16. Y. Wang, K. Cui, B. Fang, F. Wang, Cost-effective fabrication of modified palygorskite-reinforced rigid polyurethane foam nanocomposites. Nanomaterials **12**(4), 609 (2022)

17. E. Akdoğan, M. Erdem, M.E. Üreyen, M.O. Kaya, Rigid polyurethane foams with halogen-free flame retardants: thermal insulation, mechanical, and flame retardant properties. J. Appl. Polym. Sci. **137**(1) (2019)

18. R.d.Á. Delucis, W.L.E. Magalhães, C.L. Petzhold, S.C. Amico, Thermal and combustion features of rigid polyurethane biofoams filled with four forest-based wastes. Polym. Compos. **39**(S3) (2018)

19. J. Wu, Y. He, Z. Yu, Failure mechanism of rigid polyurethane foam under high temperature vibration condition by experimental and finite element method. J. Appl. Polym. Sci. **137**(6) (2019)

20. Y. Hu, Z. Zhou, S. Li, D. Yang, Z. Shui, Y. Hou, Flame retarded rigid polyurethane foams composites modified by aluminum diethylphosphinate and expanded graphite. Front. Mater. **7** (2021)

21. N. Kamairudin, L.C. Abdullah, S.S. Hoong, D.R.A. Biak, H. Ariffin, Preparation and effect of methyl-oleate-based polyol on the properties of rigid polyurethane foams as potential thermal insulation material. Polymers **15**(14), 3028 (2023)

22. S. Adnan et al., Low density rigid polyurethane foam incorporated with renewable polyol as sustainable thermal insulation material. J. Cell. Plast. **58**(3), 485–503 (2022)

23. E. Kabakci, G. Sayer, E. Suvacı, O. Uysal, İ. Güler, M.O. Kaya, Processing-structure-property relationship in rigid polyurethane foams. J. Appl. Polym. Sci. **134**(21) (2017)

24. P. Acuña, Z. Li, M. Santiago-Calvo, F. Villafañe, M. Rodríguez-Pérez, D. Wang, Influence of the characteristics of expandable graphite on the morphology, thermal properties, fire behaviour and compression performance of a rigid polyurethane foam. Polymers **11**(1), 168 (2019)

25. Y. Chen, Y. Luo, X. Guo, L. Chen, T. Xu, D. Jia, Structure and flame-retardant actions of rigid polyurethane foams with expandable graphite. Polymers **11**(4), 686 (2019)

26. S. Oprea, A. Joga, B. Zorlescu, V. Oprea, Effect of the hard segment structure on properties of resorcinol derivatives-based polyurethane elastomers. High Perform. Polym. **26**(8), 859–866 (2014)

27. S. Sang, Y. Li, K. Wang, T. Jia-ling, Application of blocked isocyanate in preparation of polyurethane(urea) elastomers. J. Appl. Polym. Sci. **138**(24) (2021)

28. Y. Ma et al., Wide temperature range damping polyurethane elastomer based on dynamic disulfide bonds. J. Appl. Polym. Sci. **139**(2) (2021)

29. A. Magiera, M. Kuźnia, A. Błoniarz, A. Magdziarz, Rigid polyurethane foams modified with soybean-husk-derived ash as potential insulating materials. Processes **11**(12), 3416 (2023)

30. M. Hoseinabadi, M. Naderi, M. Najafi, S. Motahari, M. Shokri, A study of rigid polyurethane foams: the effect of synthesized polyols and nanoporous graphene. J. Appl. Polym. Sci. **134**(260) (2017)
31. A. Hürkamp et al., Integrated computational product and production engineering for multi-material lightweight structures. Int. J. Adv. Manuf. Technol. **110**(9–10), 2551–2571 (2020)
32. W. Zhang, M. Tong, F. Xing, H. Zheng, C. Zhu, Enhanced flame-retardant performance of rigid polyurethane foam by using aptes-mmt and ath mixed intumescent coatings. J. Appl. Polym. Sci. **141**(4) (2023)
33. H. Singh, Rigid polyurethane foam: a versatile energy efficient material. Key Eng. Mater. **678**, 88–98 (2016)
34. H. He et al., Chemical recycling of waste polyurethane foams: efficient acidolysis under the catalysis of zinc acetate. ACS Sustain. Chem. Eng. **11**(14), 5515–5523 (2023)
35. J.B. Dunn, L. Gaines, J.L. Sullivan, M.Q. Wang, 'Impact of recycling on cradle-to-gate energy consumption and greenhouse gas emissions of automotive lithium-ion batteries. Environ. Sci. Technol. **46**(22), 12704–12710 (2012)
36. A. Li, D. Yang, H.N. Li, C.L. Jiang, L. Ji, Flame-retardant and mechanical properties of rigid polyurethane foam/mrp/mg(oh)2/gf/hgb composites. J. Appl. Polym. Sci. **135**(31) (2018)
37. B. Zhang et al., Bio-based trivalent phytate: a novel strategy for enhancing fire performance of rigid polyurethane foam composites. J. Renew. Mater. **10**(5), 1201–1220 (2022)
38. U. Soykan, Ş Kaya, Role of hemp fiber addition on thermal stability, heat insulation, air permeability and cellular structural features of rigid polyurethane foam. Cell. Polym. **42**(2), 88–104 (2023)
39. S. Wang, L. Qian, F. Xin, The synergistic flame-retardant behaviors of pentaerythritol phosphate and expandable graphite in rigid polyurethane foams. Polym. Compos. **39**(2), 329–336 (2016)
40. S. Yang et al., Fire retarded polyurethane foam composites based on steel slag/ammonium polyphosphate system: a novel strategy for utilization of metallurgical solid waste. Polym. Adv. Technol. **33**(1), 452–463 (2021)
41. A. Akar, B. Değirmenci, N. Köken, Fire-retardant and smoke-suppressant rigid polyurethane foam composites. Pigment Resin Technol. **52**(2), 237–245 (2022)